WARBIRDTECH SERIES

VOLUME 21

AF709550

BOEING / BAe HARRIER

By Dennis R. Jenkins

Specialty Press
PUBLISHERS AND WHOLESALERS

Copyright © 1998 Dennis R. Jenkins

Published by
Specialty Press Publishers and Wholesalers
11481 Kost Dam Road
North Branch, MN 55056
United States of America
(612) 583-3239

Distributed in the UK and Europe by
Airlife Publishing Ltd.
101 Longden Road
Shrewsbury
SY3 9EB
England

ISBN 1-58007-014-0

All rights reserved. No part of this book may be reproduced or transmitted in any form or by any means, electronic or mechanical including photocopying, recording or by any information storage and retrieval system, without permission from the Publisher in writing.

Material contained in this book is intended for historical and entertainment value only, and is not to be construed as usable for aircraft or component restoration, maintenance or use.

Designed by Dennis R. Jenkins

Printed in the United States of America

Table of Contents

The Boeing / BAe Harrier

Preface

A Word From the Author

The Harrier is an interesting aircraft. After it first entered service it was a frequent visitor to various airshows around the world, and it would typically introduce itself by hovering in front of the crowd and taking a bow, slowly dipping its nose about 15 degrees. The crowds loved it. Unfortunately, several aircraft picked that moment to have engine or control problems, giving the crowds much more of a show than they had expected, with fatal results in several instances.

And therein lies the Harrier's Achilles Heel. Although not really a dangerous aircraft, it has nevertheless amassed the worst safety record of any current Western fighter – its accident rate per flight hour is almost three times that of the F-15 or F-16. This is not necessarily the fault of the technology. The Harrier operates in a more critical flight regime than other fighters. Takeoff and landing are usually the most dangerous times for any aircraft, and the Harrier spends a large percentage of its time in this regime. This is the nature of its mission – close air support – especially in US Marine Corps service. It supports the ground forces from locations essentially collocated with them. It takes off, pounds the enemy, and lands – quickly. It generates more sorties than other fighters simply because it is closer to the action.

The Harrier is also one of the few examples of a foreign aircraft that has been successful in American military service. Even though most current Harriers were largely manufactured by McDonnell Douglas (now part of Boeing), it is still very much a product of British design. The Americans provided a great deal of refinement to both the airframe and engine, but the credit for the concept resides squarely with Hawker Siddeley (now part of British Aerospace), and Bristol (now part of Rolls-Royce Aircraft Engines).

The Harrier has been flying in one form or another for almost 40 years, and it is still the only truly workable fixed-wing V/STOL fighter in service. Six countries operate Harriers – some old, some new – but all unique in the world of aviation. Surprisingly, the Harrier seems almost singularly responsible for the emergence of aircraft carriers in navies other than the United States and France.

Harrier Loss Rate per Flight Hour Comparison

# of Flight Hours	90,000	213,000
Harrier (USMC and RAF)	25	50
A-10 (USAF)	8	17
F-15 (USAF only)	4	15
F-16 (USAF only)	10	30

The United States is attempting to develop a follow-on V/STOL aircraft in the Joint Strike Fighter (JSF), currently expected to enter service in 2010. That gives the Harrier at least ten more years of service since it is unlikely that its operators will give it up without a direct replacement.

There are several good books about the Harrier on the market. Probably the best is *Harrier: The Vertical Reality* by Roy Braybrook and published by the Royal Air Force Benevolent Fund Enterprises. An AIAA case study by John W. Fozard, who was the chief designer of the Harrier, provides a fascinating insight into the technical problems associated with designing the world's first successful V/STOL fighter aircraft.

My thanks go to my good friends Mick Roth and Terry Panopalis for all of their help. A great deal of credit must also go to Kev Darling at Big Bird Aviation in the UK for supplying a great deal of research material on British Harriers. Kev and Nick Challoner (www.airspeed.ndirect.co.uk) furnished many of the European Harrier photos presented here. All of their assistance is greatly appreciated.

Dennis R. Jenkins
Cape Canaveral, FL
October 1998

1 Vertical Flight

An Engineer's Dream

Many concepts for vertical takeoff rotary-wing vehicles have been presented since Leonardo da Vinci first sketched a helicopter in 1483, but the idea did not bear fruit until a coaxial-rotor helicopter constructed by Louis Breguet and Rene Dorand in France made its first flight in 1935. With the perfection of the single-rotor type by Igor Sikorsky in the United States during 1939-41, the helicopter finally became a practical aircraft capable of a variety of missions. It was quickly adopted by both military and civilian operators in almost every country in the world.

World War II also brought the first serious thought of using vertical takeoff for fixed-wing aircraft. Previously there had not been much reason to pursue the concept since the aircraft of the day required relatively short takeoff runs on unprepared grass strips. The Luftwaffe was the first to show any serious interest in the concept, mainly because their latest aircraft (such as the jet-powered Me 262) were beginning to require longer takeoff runs and Allied bombing was taking a heavy toll on available runways.

The simplest approach to vertical takeoff was simply to install a powerful engine and sit the aircraft on its tail, much like a rocket. This was exactly what was done with the Bachem Ba-349-A1 Natter, which was a point-defense fighter intended as a last-ditch interceptor against Allied bombers. Work on the aircraft began in late 1944, and by the end of the war there were 200 on order for the SS and Luftwaffe. The Natter was a small single-seat rocket-powered interceptor that would be launched vertically from a ramp and flown almost straight up under direction from a ground-based radar site. As soon as enemy bombers were sighted by the pilot, he would attack with two-dozen unguided 73-mm rockets. After the attack, the Natter pilot separated the cockpit cover and left the aircraft. Both the pilot and aircraft descended to the ground under separate parachutes.

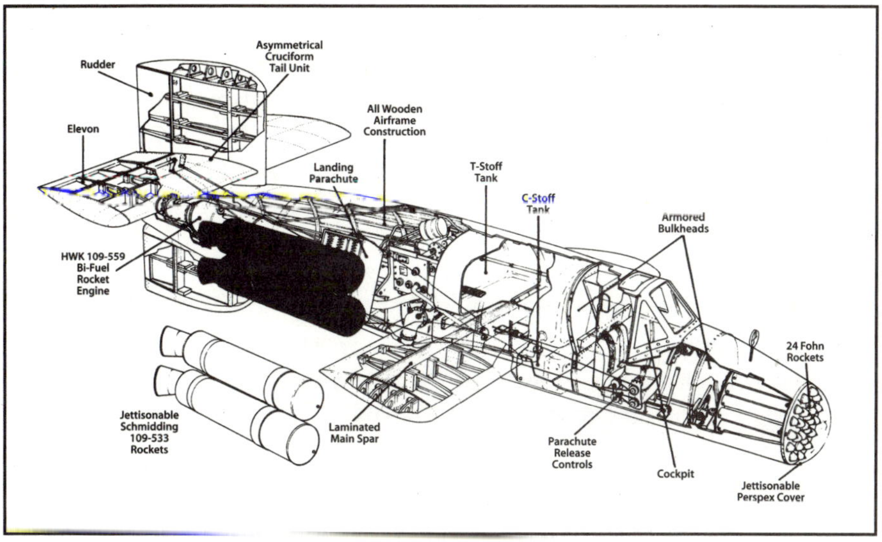

The Bachem Ba-349-A1 Natter was a last ditch effort by Nazi Germany to produce an inexpensive, high-performance, point-defense interceptor. Its single manned test flight (right) was unsuccessful. (U.S. Air Force Museum)

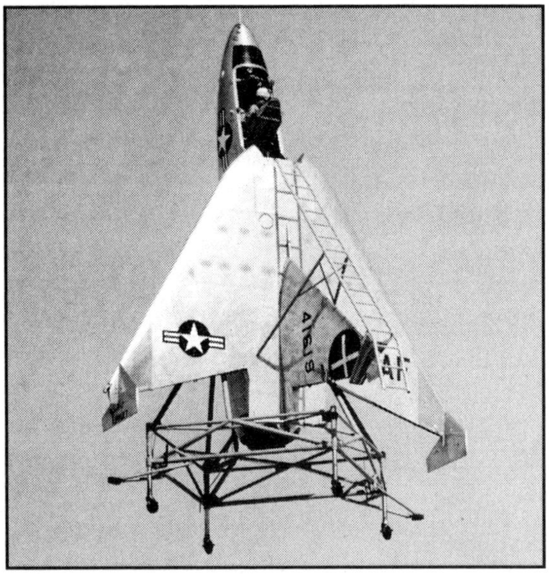

Like many early efforts, the Ryan X-13 was a tail-sitter. Although this offered some advantages from a propulsion perspective, the operational limitations were never overcome. (Teledyne Ryan)

Only a single manned test flight was conducted, and it was less than successful. After the Natter had risen approximately 500 feet the cockpit cover flew off. The pilot lost control of the aircraft, which continued to climb to approximately 5,000 feet before entering a flat spin and crashing near the launch site. The pilot, Löthar Siebert, was killed and the aircraft destroyed. No further flight tests were carried out prior to Allied forces capturing the area. Only a handful of the 200 Natters had been completed.

The next serious attempt at vertical takeoff (called vertical attitude takeoff and landing – VATOL – at the time, now known as vertical takeoff and landing – VTOL) was the Ryan X-13 Vertijet. In late 1946 Ryan Aeronautical Company began investigating the concept of using jet power (instead of rockets) to enable a vertical takeoff. Ryan originally designed the aircraft in response to a request from the US Navy, which issued a small contract to Ryan in April 1947 to allow a more detailed study into the necessary flight control systems. The flight controls on the Natter had been extremely crude, consisting primarily of several expendable vanes located in the rocket exhaust (much like those used on the V-2) that allowed some amount of thrust vectoring during vertical flight until sufficient airspeed was gained for the aerodynamic controls to become effective.

Ryan decided to pursue what are now termed "reaction controls" as the primary means of maintaining aircraft stability during hover and nonaerodynamic flight. This concept, common on most spacecraft, uses small thrusters at various locations on the vehicle to provide attitude control. These thrusters can use separate engines, compressed air, or deflected engine air depending upon many variables considered during vehicle design. The Navy contract allowed Ryan to build an outdoor test stand to "fly" full-scale aircraft during control system evaluations. It also allowed Ryan to build a flying testbed to evaluate various thrusters and flight control systems. On 20 October 1950 the Ryan test vehicle flew for the first time under three-axis control, although it was still suspended by a gimbal arrangement from the test stand. On 31 May 1951 Ryan conducted the first free-flight hover by a jet-powered vehicle.

For a variety of reasons, the Navy did not further pursue the Ryan Vertijet (a full history of the project can be found in Jay Miller's excellent reference, *The X-Planes*). Luckily, the Air Force decided to continue the experiments and contracted to construct two Ryan Model 69 (X-13) research aircraft. On 28 May 1956 the X-13 completed its first test hop in vertical mode. This was a very conservative test; the aircraft reached an altitude of less than 50 feet and a horizontal speed of only 30 mph. Progress was relatively quick, and sufficient confidence had been gained by 30 July 1957 to conduct a demonstration flight in the Pentagon parking lot! The seven minute flight retraced the route flown by Orville Wright and Lt. Benjamin Foulois 48 years earlier when making the acceptance flight of the first US military aircraft.

But everybody seemed to lose interest in the little X-13 shortly after this, and by early 1958 the test program had been largely terminated. Both aircraft were eventually retired (the first to the Smithsonian Institution; the second to the Air Force Museum), but they had successfully demonstrated that a jet aircraft could be controlled during vertical takeoff and nonaerodynamic hover. Perhaps more importantly, the aircraft had proven that VTOL flight on jet thrust alone was both technically feasible and practical.

Other contractors had also been thinking about VTOL concepts. From the mid-1940s through the 1950s one of the most innovative aerospace contractors was Bell Aircraft Company. As early as 1941 Bell had filed a patent application for a VTOL tail-sitter powered by a single reciprocating engine driving contrarotating propellers. The patent was finally granted in August 1945. By that time Bell had conducted preliminary evaluations of a similar aircraft powered by the new turbojet engines but had concluded that the engines were not sufficiently powerful for the task.

By 1950, Bell had completed more detailed studies of VTOL aircraft and they came to a remarkable conclusion – the aircraft should be horizontal-attitude designs instead of the previous vertical-attitude designs (i.e., tail sitters). A Navy competition conducted about this time reached other conclusions, and awarded contracts for the Lockheed XFV-1 and Convair XFY-1 fighter prototypes – both were propeller-driven tail sitters. Bell had concluded that the control problems, along with orientation difficulties encountered by the pilot in the vertical mode, were too great to overcome in an operational aircraft. This was confirmed during test flight of the XFV-1 and XFY-1.

Bell continued to investigate the concept of a horizontal-attitude VTOL aircraft, and during April 1952 was awarded an Air Force feasibility study contract. Although many of the control problems during hover were the same, at least the pilot was in a more conventional attitude and the flight controls only had to operate in a single mode. A year later the company built the Model 65 Air Test Vehicle (ATV) to obtain some practical experience with their concepts. The ATV (civil registration N1105V) was powered by two Fairchild J44 jet engines that had originally been designed for various US cruise missiles. The engines were mounted beneath the wings and could rotate 90° from vertical to horizontal, allowing the aircraft to transition from vertical to normal wingborne flight. A compressed air system was installed with a single outlet at each wingtip and two in the tail (pitch and yaw) for control during nonaerodynamic flight.

The ATV program was terminated in early 1955 after 4.5 flight hours had been accumulated since the cruise missile engines had only been designed for a nominal 5 hour life. Although the aircraft never completed a full transition from vertical to horizontal flight, it provided a wealth of information on VTOL flight.

The limited testing of the ATV had demonstrated the concept was practical, and during January 1955 Bell submitted four reports recommending the development of a VTOL day fighter The Air Force was sufficiently interested to fund further research and development, eventually leading to the Bell D-188 design. A full-scale mockup of this aircraft was built under a tentative XF-109 designation (although there is a great amount of debate over whether this was a truly official designation). For unspecified reasons the project was terminated before a real aircraft was built.

The Air Force, however, was still interested in the VTOL concept, and in July 1955 awarded Bell a contract for a single Model 68 test aircraft under the X-14 designation. This vehicle used the principle of deflecting thrust rather than rotating the entire engine. Two rotating nozzles were used to deflect the thrust either downward or to a position approximately 21° below the horizontal. The decision not to make the nozzles rotate all

The Lockheed XFV-1 (left) and Convair XFY-1 (right) were both tail-sitter fighter prototypes powered by the 6,825 hp Allison XT40-A-16 turboprop engine. During the early 1950s this was thought to be the route to VTOL flight. The orientation problems experienced by the pilot during landing resulted in both projects being cancelled. (U.S. Navy)

The Bell X-14A was the first successful demonstration of deflected (vectored) thrust. Its predecessor, the Bell Model 65 Air Test Vehicle (insert) demonstrated that hovering on jet power alone was practical, and helped develop control laws that were later used in the X-14. (NASA)

the way to a horizontal position was made based on the engineering complexity, and the fact that high-speed horizontal flight was not an objective of the program.

Hover tests began on 17 February 1957, and partial transitions to horizontal had been demonstrated by June. The first full transition took place on 24 May 1958, with the complete maneuver taking somewhat less than 30 seconds to complete. During October 1959 the aircraft was turned over to NASA at the Ames Research Center. NASA test pilots noticed was that the aircraft was seriously underpowered with its two Armstrong-Siddeley Viper 8 engines, each rated at only 1,900 lbf. A pair of 2,680 lbf General Electric J85-GE-5 engines were located and installed in the aircraft, which was redesignated X-14A.

The X-14 verified that vectored (deflected) thrust was a viable concept for a VTOL aircraft. However, the actual technique used by the X-14 had some serious drawbacks. Since the thrust was deflected only at the exhaust nozzle of the engine, this meant the engine had to be located very far forward so that the thrust was at roughly the aircraft's center of gravity. This location was not desirable in a combat aircraft since the radar, avionics, and pilot were traditionally located in this position.

A variety of other VTOL concepts were also investigated during the 1950s. Dedicated lift engines buried in the wings or fuselage, such as those used in the Short S.C.1, Yak-38, and Mirage III-V demonstrated their own drawbacks, namely a significant increase in weight and an inefficient use of internal volume. A unique set of lift fans, driven from the primary jet engine, was used for lift in the Ryan XV-5A Vertifan, but this added mechanical complexity and took valuable wing space that is usually used for fuel or weapons. Although both of these concepts were tried in numerous aircraft, and in the case of the Yak and Mirage actually committed to limited production, both were considered unsuccessful, as well as inelegant from an engineering perspective.

Interestingly, both techniques are making a comeback at the beginning of the 21st Century. The Lockheed Martin X-32 JSF demonstrator uses an engine-driven lift-fan, while a dedicated lift engine is included in the competing Boeing X-35 design. It remains to be seen if advances in technology will allow these to be more successful than their predecessors.

But in the early 1960s it appeared that a new concept was needed in order to produce a truly workable VTOL aircraft. About the same time the emphasis on "vertical" was relaxed somewhat, and the term V/STOL (vertical/short takeoff and landing) became the accepted description of what was desired from any future aircraft.

The Mirage III-V was the first V/STOL aircraft capable of speeds in excess of Mach 2. Dedicated lift-engines in the fuselage (note the open air intake and exhaust doors) provided vertical thrust, while a conventional jet engine provided power for wingborne flight. (Dassault)

2 Innovative Engineering

The Pegasus Vectored Thrust Engine

The engineering concept that would eventually allow the first truly successful V/STOL fighter was largely conceived by a French engineer named Michel Henry Marie Joseph Wibault during the early 1950s. He envisioned using a high-powered turboshaft engine, such as those normally used in helicopters, to drive large centrifugal compressors mounted on the side of an aircraft's fuselage. The airflow could be vectored simply by turning the compressor casings, providing vertical lifting thrust, normal propulsive thrust, or at intermediate settings, some combination to aid in transition from vertical to horizontal flight. Wibault's first design was based on the 8,000 hp Bristol BE.25 Orion turboprop, but he quickly discovered that four (instead of two) nozzles would be necessary to achieve the balance needed for vertical flight. This added a great deal of weight and complexity to the design since it dictated the use of a second gearbox. Nevertheless, Wibault proceeded with his "Ground Attack Gyroptére," and presented the design in 1956 to both the French Ministry of Defense and the US Department of Defense (DoD), both without success.

In early 1956 the design was presented to NATO's Mutual Weapons Development Agency (MWDA) and found some minor support from US Army Col. John Driscoll. Understanding the merit of a workable V/STOL fighter to provide close air support to troops at the front lines, Driscoll sent the Wibault design to the Bristol Aero Engines Company for evaluation. Engineers at Bristol also saw potential in the design, and set about improving the concept.

Although the Wibault design was innovative, its mechanical complexity worried engineers at Bristol. At the time the company was involved in developing the Olympus 21 engine, something that could loosely be called a turbofan today. In this design the fan section was separated from the basic engine and driven by a long drive shaft. Separate intakes and exhausts were used for the fan and engine, so the fan did not act as a compressor stage for the basic engine. Gordon Lewis, the Bristol engineer assigned to evaluate the Wibault concept, suggested that the four centrifugal compressors should be replaced by a single reduction-geared axial-flow fan, based on the one being developed for the Olympus 21, mounted coaxial with the Orion engine. The fan would exhaust through a rotatable elbow nozzle on either side of the fan casing, while the Orpheus exhaust would be discharged in a conventional rearward jet pipe.

The revised powerplant was designated BE.48 and presented to Wibault in early 1957. Much to the surprise of Lewis, Wibault liked the concept and began modifying his aircraft design accordingly. Further investigations revealed that the reduction gearbox between the fan section and engine still added too much complexity and weight to the design, so Bristol engineers undertook adapting the vectored thrust fan section to the single-shaft Orpheus engine. The change from the high-pressure Orion to the low-pressure Orpheus resulted in higher specific fuel consumption, but the powerplant itself was much lighter, and the engineers thought it would be more robust.

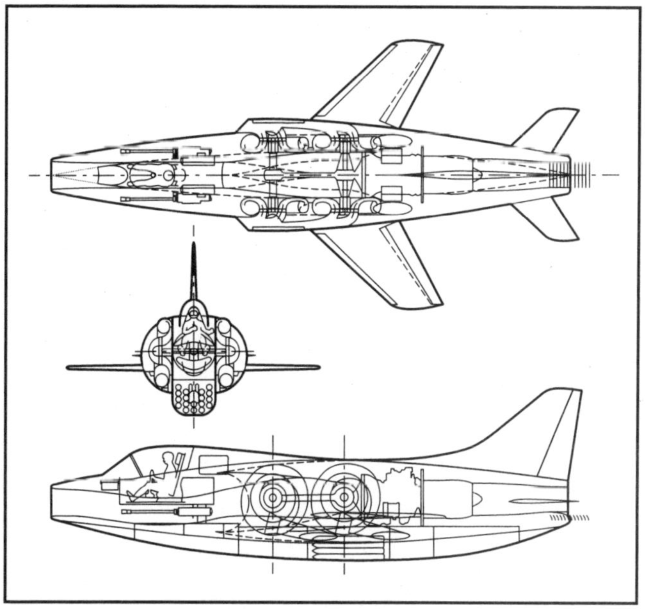

Wibault's original Gyroptére had four rotating nozzles and a normal straight-through jet exhaust in the rear. Armament consisted of two cannon and 18 unguided rockets in a retractable belly pod. (British Aerospace)

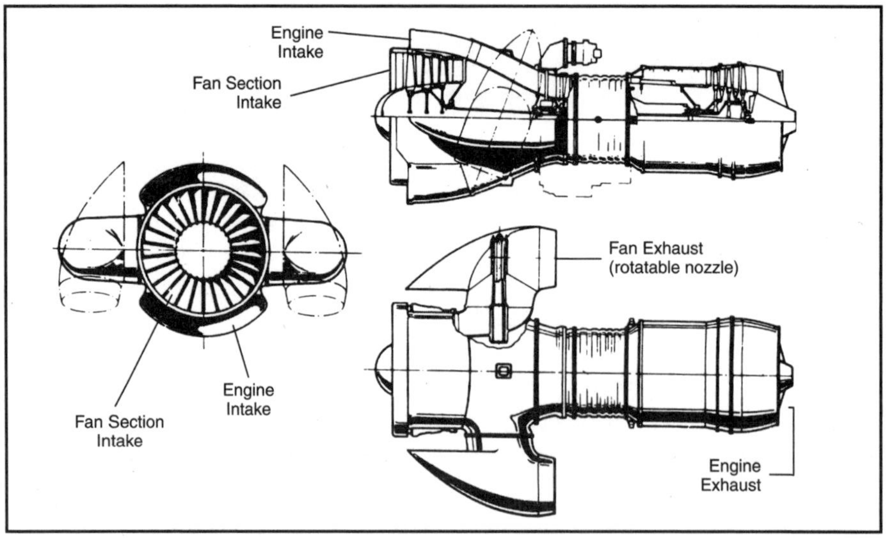

The BE.48 had a large fan section which was equipped with two rotating nozzles to discharge compressed air, while the normal turbojet engine exhaust discharged through a jet pipe fitted with a diffuser that allowed the thrust to be angled downward. (Rolls-Royce Aircraft Engines)

The fan section would provide a mass airflow equal to 3.5 times the air ingested by the Orpheus itself, resulting in approximately 75% of the static thrust being delivered by the two rotatable nozzles. Surprisingly, the intake passages for the fan and engine were separate, mainly because engineers did not want to figure out how to accommodate such different air flow requirements.

Two British patents were granted for the new engine concept, naming Wibault and Lewis as joint inventors. Various techniques were described in the patent application for changing the "balance" of the engine by diverting fan air into the main engine to create a true turbofan (thus changing the ratio of thrust between the front and rear nozzles), and using contrarotating engine/fan spools to minimize possible gyroscopic effects. By mid-1957 the engine had been refined as the 11,300 lbf BE.53 using a three-stage fan derived from the Olympus and an uprated Orpheus engine section.

Now that they had a seemingly workable engine, Bristol needed an aircraft manufacturer to use it. At the time, Short Brothers was designing the S.C.1 VTOL aircraft that used separate lift and propulsion engines in a concept being promoted by engine manufacturer Rolls-Royce. Nevertheless, Bristol approached Short to use the BE.53 engine concept. Although Short expressed some minor interest in the new Bristol engine, they were reluctant to abandon the Rolls concept, which was generally accepted as the only workable, if not ideal, solution to high-speed V/STOL flight.

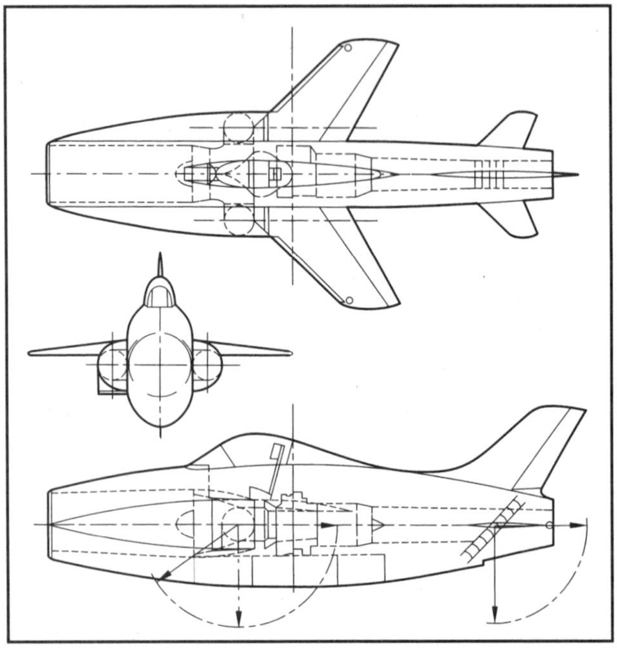

Wibault's revised Gyroptére used the BE.48 engine and showed how the front "cold nozzles" and rear diffuser allowed the thrust to be vectored downward for vertical flight. Unfortunately, engineers at Bristol could not figure out how to balance the aircraft and went back to the drawing boards. (British Aerospace)

Next to be approached was Hawker Aircraft. Actually, it appears that Sir Sydney, the director of Hawker, made the initial contact when he wrote to Dr. Hooker of Bristol after viewing films showing the X-13's first hover. Sydney wondered if more attention should be paid to this VTOL concept, but stated he had little interest in developing yet another experimental aircraft. Hooker replied by sending data on the BE.53 engine, sparking some interest at Hawker.

The BE.53 specification still used the two rotating forward nozzles and a conventional exhaust for the jet engine. The engineers at Hawker could not figure out how to build a stable aircraft around this configuration, but nevertheless drew up a preliminary design for the P.1127 "High Speed Helicopter." The Hawker team continued to investigate the concept and decided to "split" the Orpheus jet pipe and fit two additional elbow nozzles, thus permitting the entire engine thrust to be vectored. Hawker also, or so they thought, originated an idea to use contrarotating spools to minimize any possible gyro effect from the engine. Although many papers have credited this advance to the Hawker team, in fact Bristol had included this possibility on their original patent applications over a year earlier. Nevertheless, the Hawker team can be credited with actually getting some hardware built using the concept. Hawker intended to use bypass air from the compressor stage ducted to valves at the the wingtips, nose, and tail to provide directional control during low speed flight.

The Short Brothers S.C.1 used dedicated Rolls-Royce lift engines to allow vertical flight. Although it was guaranteed to work, this design made a very inefficient use of fuselage volume, and added a great deal of mechanical complexity. (Rolls-Royce Aircraft Engines)

The Hawker team also contributed other major changes to the engine specifications – one of them combined the intakes for the fan and engine stages into a single entry, greatly simplifying the aerodynamics of the aircraft.

By late 1957 the P.1127 had been modified to incorporate the new engine features, and bore little resemblance to the original P.1127 design. The basic features of what would become the Harrier were beginning to emerge, and even at this early stage the design was intended as a military aircraft. Space and weight were reserved for military equipment, external weapon stations were located under the wings, and full military load factors were used in designing the structure. Estimates included a vertical takeoff weight of 8,500 pounds, which happened to correspond to a clean aircraft with full internal fuel. An additional 2,000 pounds of weapons could be carried externally (there were no internal weapons), but this would require a 600-foot takeoff roll.

Since it was the NATO Mutual Weapons Development Agency that had originally introduced Bristol to the Wibault engine, it was fitting that the MWDA was the only official entity to express any interest in the P.1127 concept. In June 1958 the MWDA (essentially the US Department of Defense) agreed to fund 75% towards building a small number of BE.53 engines. Bristol provided the remaining 25% as a private venture since no funding was available from the British government. At the same time the MWDA began writing a specification for a lightweight V/STOL close air support aircraft based closely on the P.1127 design. The specifica-

The secret ingredient: one of the Kestrels (this one in the U.S. Air Force Museum) shows one of the "cold nozzles" that help balance the aircraft during hover and vertical flight. Wibault first proposed the idea of rotating the cold nozzles (so named because they discharge compressed fan air, not engine exhaust). Hawker's contribution was applying similar nozzles at the back of the engine to discharge hot engine exhaust. (Chris M. Reed)

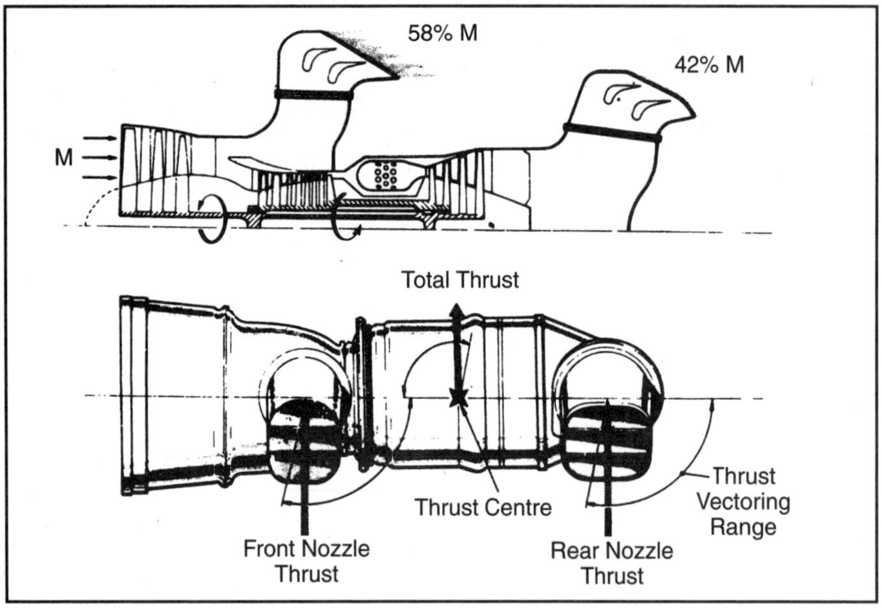

By the beginning of 1959, the Pegasus engine had taken on most of its final shape. This illustration shows 58% of total airflow being discharged from the front "cold" nozzles, with the remaining 42% coming from the rear engine exhaust nozzles. The center of thrust is roughly in between the two sets of nozzles. The thrust vectoring range could vary between slightly forward (to allow rearward flight) to completely aft for high speed flight. (Rolls-Royce Aircraft Engines)

tion called for a V/STOL replacement for the Fiat G.91 that was just beginning to enter service with many of Europe's air forces. It should be noted, however, that this specification was never published in its initial form. By the time it had made it through the NATO bureaucracy, the aircraft it demanded was much different.

This is a good point to note the history of the two primary companies involved in the Harrier. Hawker Aircraft Limited, located at Kingston-upon-Thames, was absorbed into the Hawker Siddeley Group during the late 1950s. From 1963 to 1977 they were called Hawker Siddeley Aviation, Ltd., and in 1977 they became part of the British Aerospace (BAe) conglomerate. Bristol Aero Engines Ltd. became Bristol Siddeley Engines Ltd. in 1959. In 1968 the company was absorbed into Rolls-Royce Ltd. as the Bristol Engine Division. The entire company is now called Rolls-Royce Aircraft Engines, most other parts of the famous brand having been sold off to Volkswagen and BMW.

Bristol continued developing the BE.53, helped by the MWDA funding. The ex-Olympus fan section was finally replaced by a new low-pressure transonic compressor that featured the contrarotating spools everybody was taking credit for. The first modified engine ran in September 1959 and delivered 11,300 lbf at sea level. This revised engine was named Pegasus.

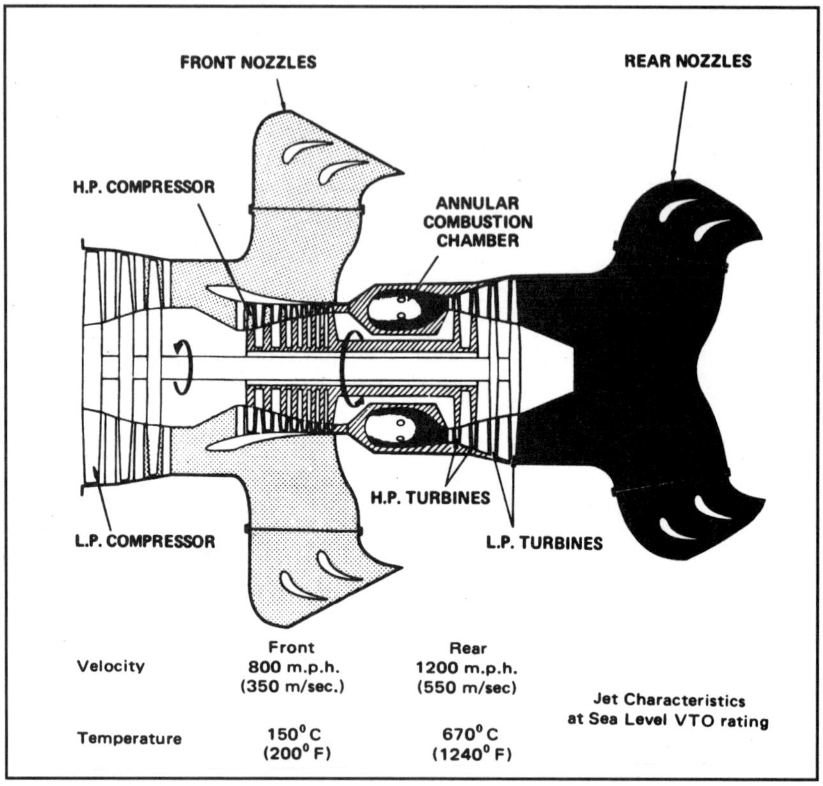

The Pegasus engine, as finally evolved, is a true turbofan, with the fan section supplying air to the forward nozzles, and also to the engine section. The black areas are "hot", while the light grey areas are "cold." Note that the spools counter-rotate, eliminating any residual torque from the engine that could affect aircraft handling. (Rolls-Royce Aircraft Engines)

P.1127 AND P.1154

Subsonic and Supersonic V/STOL for NATO

In December 1959 the NASA Langley Research Center became interested in the P.1127 design and, in a cooperative venture with Hawker, began testing various wind tunnel models. This testing revealed that the design had satisfactory flight characteristics, and in February 1960 a 1/6th-scale free flight model conducted the first full transitions from vertical to horizontal flight with no major problems. The data generated during these tests greatly encouraged the engineers working on the P.1127 project.

Further refinement by Hawker resulted in some changes to the basic aircraft. The most notable was the substitution of high-pressure bleed air for directional control in place of the low-pressure bleed previously specified. The reason for the change was that the low pressure air required ducts that were too large to fit into the wings.

Finally, in late 1959, the British Government (specifically, the Ministry of Supply – M.o.S.) issued a £75,000 contract to Hawker for two prototype P.1127s under Specification ER.204D, although engine funding was still the responsibility of Bristol and the MWDA. A single structural test airframe was also funded. The first of these prototypes (XP831) conducted its first hovering trials on 21 October 1960 at the Royal Aircraft Establishment, Bedford. To save weight during tethered tests, the aircraft had been stripped of most noncritical equipment, including the landing gear doors and the radio (a telephone was installed for the pilot to maintain communications), bringing its empty weight down to 9,243 pounds. On 19 November the aircraft conducted untethered hover tests, and the first conventional takeoff and landing occurred on 13 March 1961.

The first prototype continued to open up the flight envelope and was joined by the second prototype (XP836) in mid-1961. During this time there were several revisions to the reaction control system in an attempt to tailor it to the aircraft's handling characteristics. Several problems were encountered during flight tests, but none were considered showstoppers, and most had been expected in one form or another. Since the British military was finally beginning to show a mild interest in the design, the M.o.S. funded four development batch (DB) aircraft (XP972, 976, 980, and 984). Engines were still being funded mainly by the MWDA. The first of the DB aircraft (XP972) made its maiden flight on 5 April 1962 powered by a 12,500 lbf Pegasus 2 engine.

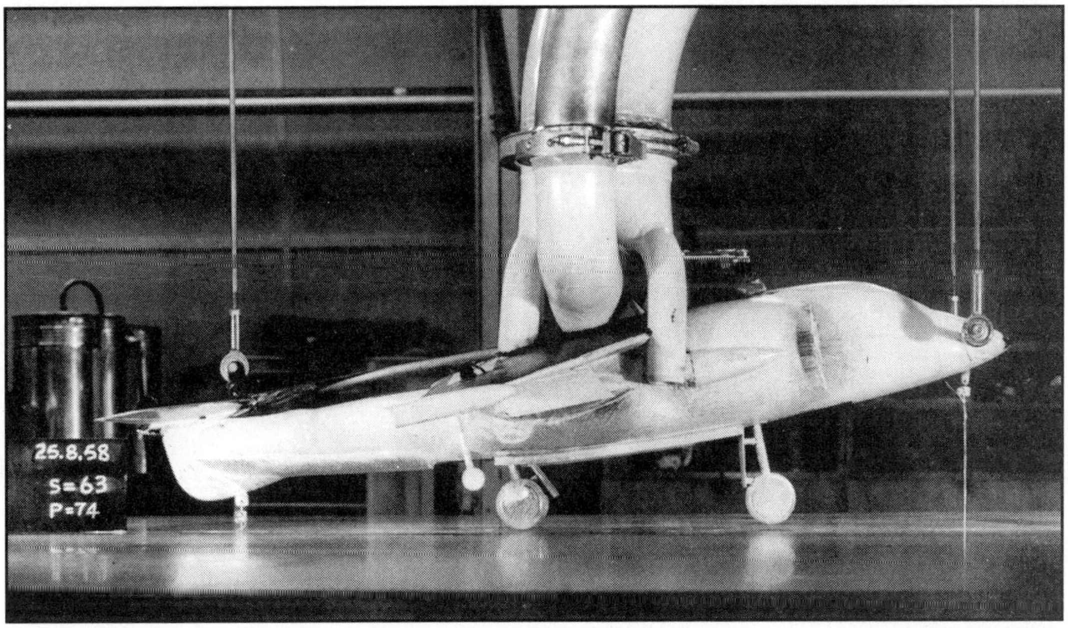

This 1/10th-scale model of the P.1127 was used for static tests of ground effect. The large pipes carried compressed air that was discharged in the same locations as the engine nozzles. The air carried colored smoke that could be photographed so that air currents while the aircraft was in ground effect could be understood. (NASA)

Flight tests with the first two prototypes had revealed some instabilities that were traced to the wing shape, and XP972 featured outrigger landing gear more smoothly faired into the wingtips. Although the revised wingtips partially cured the instability problem, Hawker engineers eventually designed a wing that abandoned the original clipped-delta shape for a more traditional swept leading and trailing edge with new fairings at the wingtip to accommodate the outrigger landing gear and reaction control valves. The new wing was ready in time to be installed on XP984. A tendency for the aircraft to pitch up at high subsonic Mach numbers resulted in XP980 being fitted with a new horizontal stabilizer with a pronounced 18° anhedral. This did not totally cure the problem, so the horizontal stabilizer was increased in span (from 12.0 to 13.75 feet) to provide more surface area and the anhedral reduced to 16 degrees.

Although progressing well, the flight tests were not without incident. On 14 December 1961 Bill Bedford was forced to eject from XP836 after the left-hand forward nozzle fell off, leaving the aircraft uncontrollable. At this point the two forward nozzles (called "cold" nozzles since combustion gases were not ejected through them) were made of fiberglass, which was rather fragile. On 30 October 1962 Hugh Merewether made a very hard landing at Tangmere in XP972 following an engine fire after a high-g maneuver starved a main bearing of oil. And during the 1963 Paris Air Show, Bill Bedford crashed in XP831 after he failed to transition from a hover to horizontal flight. It was found that a fragment of dirt had penetrated the air motors which rotated the forward nozzles, jamming them in place. Fortunately, Bedford was unhurt. Although two of the P.1127s had crashed during their trials, thankfully with no fatalities, the aircraft provided much needed confirmation that the concept of fixed-wing VTOL flight was indeed feasible. One of the survivors (XP831) was retired to the RAF Museum at Hendon.

In early 1962 the British, American, and German governments agreed to fund the procurement of nine P.1127 aircraft for use by a multinational test unit to assess V/STOL fighter operations. The last DB aircraft (XP984) that had been funded by the M.o.S. was chosen to be the prototype for the new version. It was modified with a nine-inch fuselage extension to eliminate a slight mismatch between the aircraft's

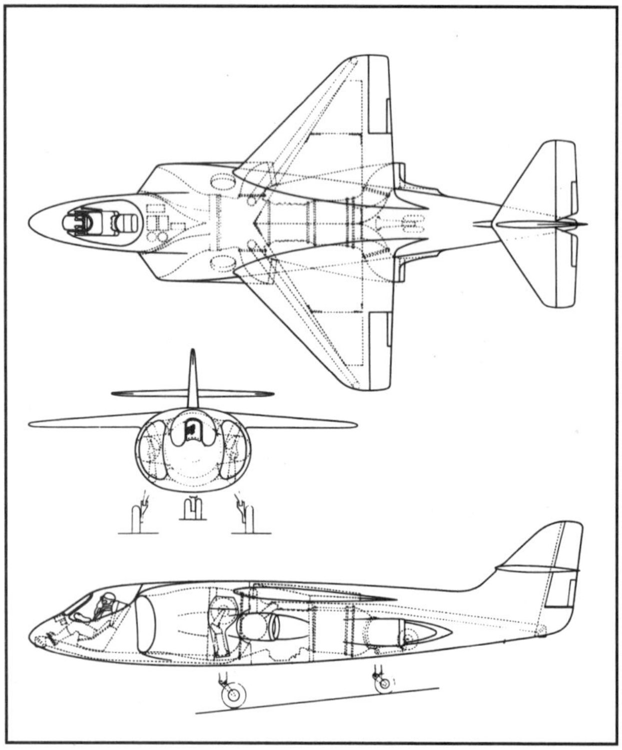

This was the P.1127 configuration during mid-1957. Most of the important aspects of the Kestrel were already in place. (British Aerospace)

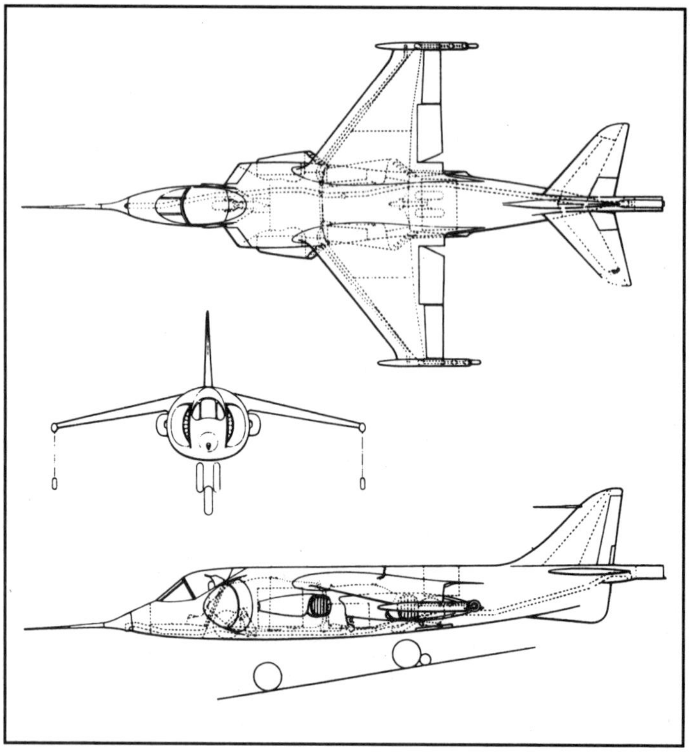

This is how the first P.1127 prototype (XP831) looked when it was rolled out in 1959. Note the wingtip pods to contain the outrigger landing gear. (British Aerospace)

The first Kestrel (XP831) during an early test flight. The black intake lip indicates this aircraft was fitted with the inflatable "lips" that were used to improve airflow during low-speed flight and hover. (British Aerospace)

center of gravity and that of the latest Pegasus 5, rated at 15,200 lbf. The new aircraft would be named Kestrel, with a British designation of F.G.A. Mark 1.

The nine Kestrels (XS688/696) were manufactured during 1964 and 1965, with the first making its maiden flight on 7 March 1964. The program got off to a bad start, however, when a US Army pilot groundlooped an aircraft on 1 April, effectively destroying it. The Tripartite Squadron officially came into being on 15 October 1964, established at RAF West Raynham. The squadron was led by RAF Wing Commander David Scimgeour, with Luftwaffe Colonel Gerhard F. Barkhorn, and US Navy Commander J.J. Tyson, Jr. as his deputies. Pilots assigned to the squadron represented the RAF, Luftwaffe, US Army, and US Air Force. The trials consisted of 1,367 sorties using the eight remaining Kestrels. A full 50% of the takeoffs were vertical, with an additional 37% considered "short" – only 23% were conventional. Over three-quarters of the landings were vertical. Operations were conducted at

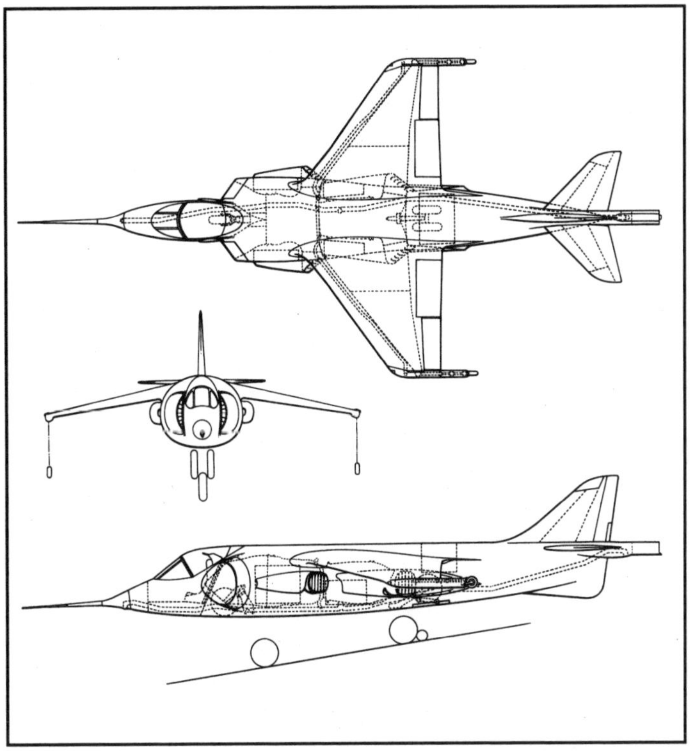

Later in its life, the P.1127 prototype (XP831) was fitted with a modified wing that significantly changed the wingtip profile. (British Aerospace)

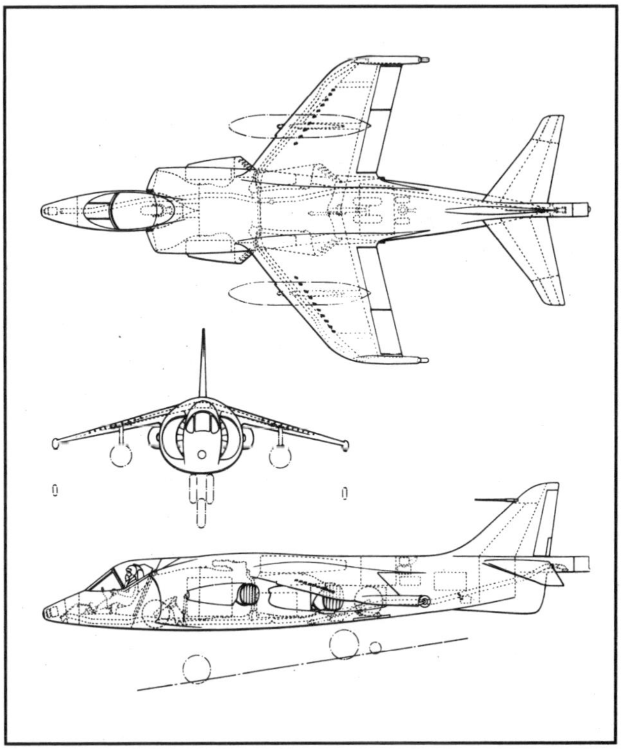

The Kestrel F.G.A. Mark 1 was an obvious derivative of the P.1127, but featured a much refined wing shape. (British Aerospace)

The second Development Batch Kestrel (XP976) shows the early wing planform with outrigger pods on the wingtips. Noteworthy is the extended ram air turbine (RAT) on top of the fuselage immediately behind the trailing edge of the wing. (British Aerospace)

seven dispersed sites using a variety of surfaces including aluminum planks and fiberglass mats, in addition to the more normal concrete, asphalt, and grass.

When the Tripartite Squadron was disbanded, it was expected that each nation would purchase an equal number of the remaining Kestrels, but Germany declined, leaving six for the Americans and two for the RAF (interesting considering it was an American pilot that wrote-off one aircraft). The six (XS688/692) assigned to the US were designated XV-6A and given serial numbers 64-18262/18267. They underwent an initial 14-week, 374-sortie evaluation at Edwards AFB. Tests were also conducted from the aircraft carrier USS *Independence* (CVA-62) and the amphibious landing ship USS *Raleigh* (LPD-1) during May 1966. Despite promising test results, the US military saw no practical application for the aircraft, and two of the aircraft were eventually transferred to NASA where they were assigned NASA numbers 520 and 521.

The two remaining RAF Kestrels were also kept busy. In July 1966 a Kestrel operated from the commando carrier HMS *Bulwark* to demonstrate compatibility with helicopter operations. Another (XS693) was bailed to Hawker Sid-

The first P.1127 conducted trials aboard HMS Ark Royal *during February 1963. Hawker chief test pilot Bill Bedford successfully proved that V/STOL aircraft could operate aboard ships.* (Terry Panopalis Collection)

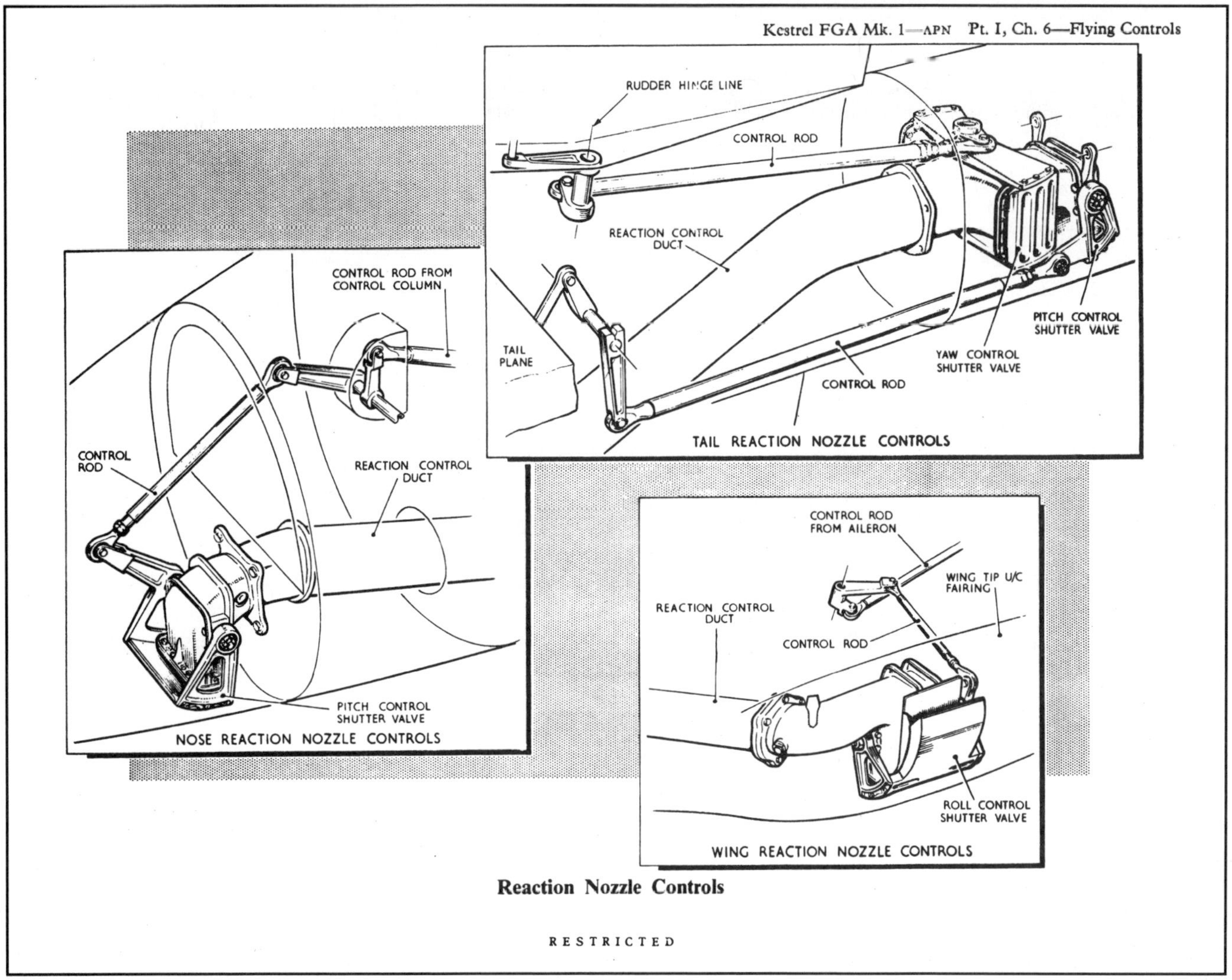

The Kestrel FGA1 reaction control system illustration from the Kestrel Pilot's Notes.

Two of the Kestrels from the Tripartite Squadron were later turned over to NASA at the Langley Research Center in Virginia. They were used for several years in various test programs. (Dave Ostrowski via the Mick Roth Collection)

The first Kestrel (XS688) is escorted by a Hunter T7 chase aircraft. Noteworthy is the insignia of the Tripartite Squadron – a combination of the normal German, British, and American insignias. Interestingly, the "fin flash" also includes colors for all three countries, even though the Americans do not normally carry a fin flash (at least, not since the early 1940s). Compare the horizontal stabilizer shape with later Harriers. (British Aerospace)

The six American Kestrels carried markings from both the US Air Force and US Army at various times during their careers. Here is the fifth aircraft wearing Air Force markings while on display at Edwards AFB in July 1967. (John Hathaway via the Mick Roth Collection)

deley as a testbed for the new Pegasus 6 engine. A set of six annular suction-relief doors around the engine air intakes was introduced to provide additional airflow during low-speed operations. It had been known for some time that the intakes needed to be modified in order to provide clean airflow at low forward air speeds. The original solution had been a set of inflatable "lips" that were activated during hovering flight to provide the desired blunt lip profile, and then deflated during forward flight to provide a streamlined low-drag shape. This solution was not considered ideal, and the suction relief doors were significantly more effective and durable. For all intents, XS693 would serve as the prototype of what eventually became the Harrier. Unfortunately, the aircraft was lost on 21 September 1967 during an accident at RAF Filton.

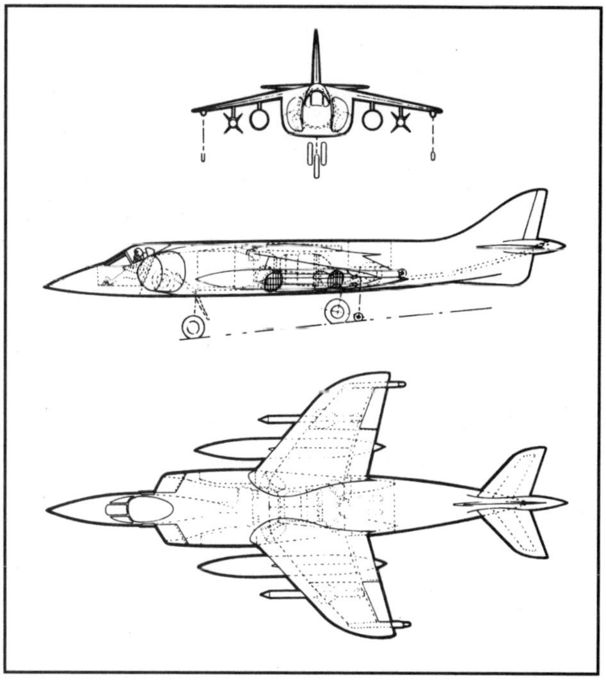

The P.1150 was the first supersonic V/STOL design, and showed an obvious relationship to the Kestrel and later Harrier. The PCB-equipped Pegasus engine would allow a top speed of over Mach 1.7 at high altitude. (British Aerospace)

The Supersonic P.1154

While the Kestrel was being developed and tested by the Tripartite Squadron, the RAF was finally becoming seriously interested in a V/STOL aircraft. But it must be remembered that during the 1960s, if an aircraft was not supersonic, it was not perceived to have military value. Therefore, Hawker began development of the P.1150, an aircraft about 50% larger than the subsonic P.1127. The engine was a version of the Pegasus featuring plenum chamber burning (PCB) where fuel was injected into the normally "cold" front nozzles and burned in the discharge air. This provided a significant increase in thrust from the front nozzles during vertical flight, which is the only time it was envisioned PCB would be used. The entire concept was much like an afterburner, except the air had never passed through an engine. A single-shock inlet design would allow speeds approaching Mach 1.7 assuming the boundary layer could be managed effectively.

During 1960 the Air Ministry began writing an Operational Requirement (O.R. 345) for an RAF version of the P.1127 fitted with a PCB-equipped version of the Pegasus rated at 18,000 lbf. Before this OR was issued to Hawker, it was superseded by NATO Requirement NBMR-3 for a very similar aircraft.

In the summer of 1961 the NBMR-3 specification was issued. It called for a supersonic V/STOL aircraft with a 250-nm radius of action flying at Mach 0.9, with a dash capability of over Mach 1.5 at high altitude. The aircraft was to be capable of clearing a 50-foot obstacle after a 500-foot take off roll with a 2,000-pound weapons load. Bidders were required to include in their proposals the cost of supplying 12 aircraft by the end of 1966.

Investigations by Hawker showed that the P.1150 could not meet these requirements. A short-lived variant featuring two dedicated lift-engines also could not meet the requirements. Interestingly, this basic design – with a non-PCB vectored trust engine squeezed between a dedicated lift engine on each end – was later built as the German VAK-191, without any great success.

Bristol therefore developed a new version of the PCB engine designated BS.100 that raised the exhaust temperature of the front nozzles to over 1,700°F, a full 500°F higher than the original proposal. This resulted in a significant boost in available thrust, and the P.1154 built around the new BS.100 was slightly larger than the original P.1150 design. The weapons system proposed for the P.1154 included a Ferranti forward looking radar for ranging and simple terrain following, coupled with a new lightweight inertial platform and associated weapon aiming and navigational computers. A moving map display was provided in the cockpit, and data was presented on a Specto (now Smiths Industries Ltd.) heads-up display (HUD). A gas turbine starter doubled as an auxiliary power unit for austere field operations.

The P.1154 proposal was submitted to NATO during January 1962, along with proposals from several other European and US manufacturers. Three months later the P.1154 was the clear choice of the multinational technical assessment team, although the Dassault Mirage III-V (which used two dedicated lift engines) was also endorsed. However, NATO did not buy aircraft (the E-3 AWACS purchased much later was the only aircraft actually procured by NATO as an organization). It was up to each individual nation to procure its own aircraft, hopefully based on NATO's technical recommendation. In this case, however, politics played a larger role than technical competence, so the Germans funded further development of the VAK-191, the French funded the Mirage III-V, the British pursued the P.1154, and the Americans watched.

For 18 months design studies were conducted of versions of the P.1154 for the Royal Air Force and Royal Navy. The process resembled what the Americans were going through with the F-111 – each service had its own requirements, and they were often at odds. The RAF wanted a single-seat, low-level strike aircraft with a secondary supersonic intercept capability. The RN wanted a two-seat, all-weather shipboard interceptor with a secondary low-level strike capability. Officially, three separate versions of the aircraft were submitted in turn, and each was rejected. Unofficially, many other variations were looked at, including one for the RN that used two Rolls-Royce Spey engines equipped with rotating side nozzles and PCB.

One final attempt was made to define a common BS.100-powered version of the P.1154, and both services rejected it as too heavily compromised to successfully accomplish either of their requirements. The joint project was officially cancelled on 2 February 1965. The RN finally received permission to purchase Spey-powered McDonnell F-4 Phantom IIs. The RAF fell back to the P.1154 version that most closely met their perceived needs and initiated detailed development under the designation P.1154(RAF).

The effort was short-lived. When the Labour party came to power in 1965 the P.1154 project was cancelled. The first development aircraft was over 30% complete at the time, and the BS.100 engine had been successfully run at over 30,000 lbf. The RAF followed the RN's lead and purchased Spey-powered F-4s.

The P.1154(RAF) (top) and P.1154(RN) designs before the project was cancelled. The first development aircraft was over 30% complete when construction was halted. The airframe was later scrapped. Development of the PCB-equipped Pegasus continued at a reduced level. (British Aerospace via the Terry Panopalis Collection)

THE HARRIER IS BORN

4

THE KESTREL SOARS TO NEW HEIGHTS

When the P.1154(RAF) was cancelled, the RAF was allowed to pursue an advanced version of the Kestrel for use as a V/STOL close support fighter. This marked the first fixed-wing "flat-riser" ordered into production destined for operational service. It was felt that the experience gained from the Kestrels would allow the normal prototype phase to be skipped, and six preproduction P.1127(RAF) aircraft (XV276/281) were ordered in late 1965. These aircraft were, in fact, the remainder of the original Kestrel order, which had been for 15 aircraft – nine for the Tripartite Squadron, plus options for six future development aircraft. These would be named Harrier in 1967.

Even as the first engineering drawings of the P.1127(RAF) were being issued in 1965, the engineers at Bristol were busy improving the Pegasus engine. A new fan featured transonic blades and more airflow capacity, and was first introduced on the 19,000 lbf Pegasus 6, which was known as the Pegasus Mk 101 when it finally entered RAF service. But the engineers were not finished. A new Pegasus 11 was being developed, although it was evident that the existing Harrier intakes could not supply sufficient airflow to allow the engine to develop full power. Therefore, in late 1966 a decision was made to slightly enlarge the Harrier air intakes, and to increase the number of suction-relief (by this time called "blow-in" doors) from six to eight per side. The cowl slot formed by the doors was sealed by freely-hinged doors on the outside of the inlet; the coupled inner doors used on the Kestrels being eliminated as unnecessarily complicated.

It was too late to change the configuration of the first five preproduction aircraft, so the new inlets did not make an appearance until the last preproduction machine. At least two of the earlier aircraft were retrofitted with the new intakes during their careers. Unfortunately, the Pegasus 11 engine was not progressing as fast as the airframe, and in late 1968 a decision was made to produce an interim engine called Pegasus 10. Known as the Pegasus Mk 102 in RAF service, this was essentially a Pegasus 6 that was run over its rated speed and developed 20,000 lbf. The newer 21,500 lbf. Pegasus 11, designated Mk 103 by the RAF, would be retrofitted at some point in the future. It should be noted that the increase in thrust available from the Mks 102 and 103 engines was only available for short periods during vertical flight – during wing-borne flight (nozzles aft) their rating was essentially the same as the Mk 101.

The Pegasus Mk 102 did have some problems, most of them related to

The normal configuration for Harrier GR1s was a pair of Aden cannon pods on the lower fuselage and extra fuel tanks on the inboard wing pylons. This aircraft (XV741) is shown in January 1974. (Mick Roth Collection)

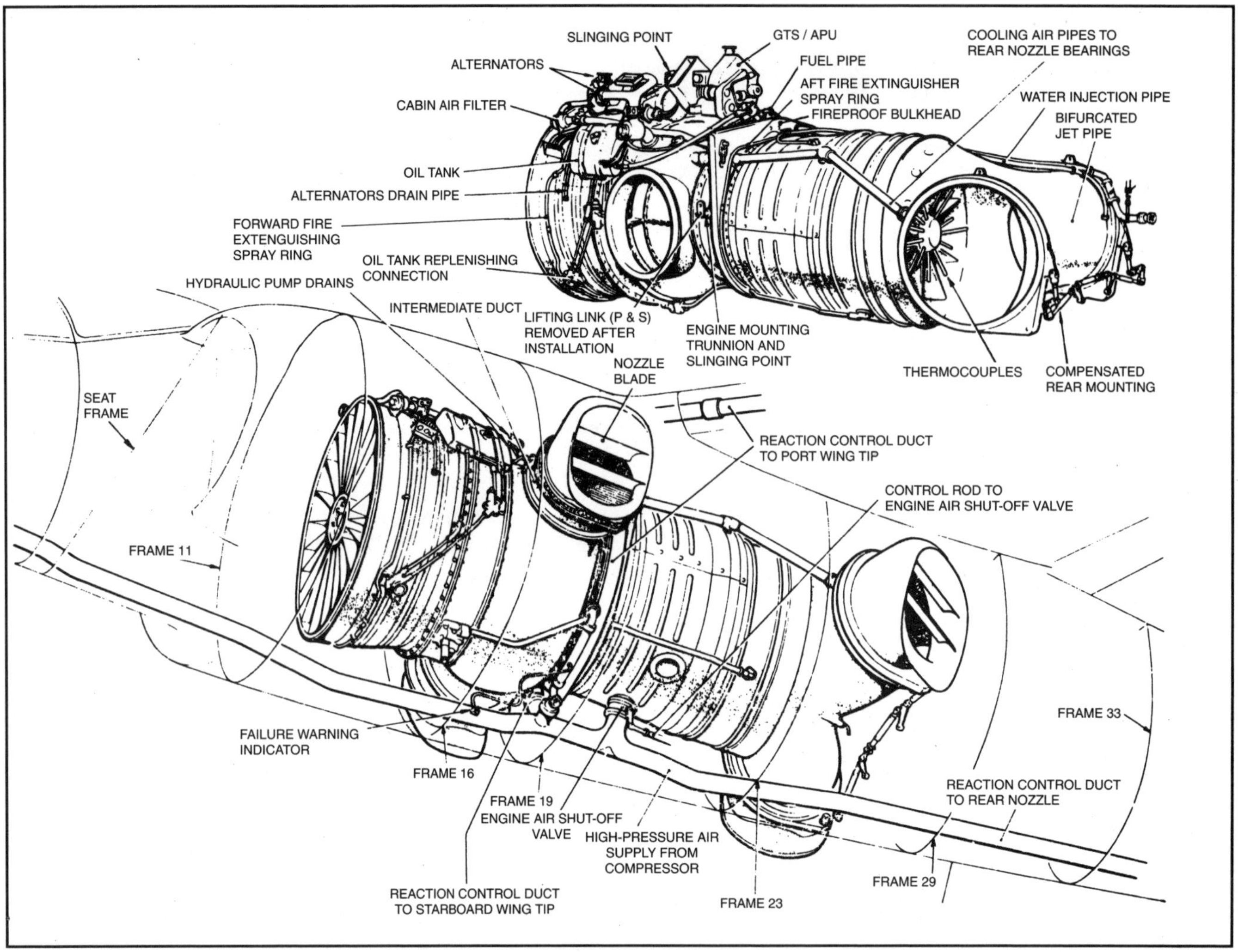

The Pegasus engine took up most of the center fuselage of the Harrier. Due to the size of the engine in relation to the airframe, and the unique configuration of the four exhaust nozzles, Hawker decided upon an unorthodox method of installing the engine – the entire wing lifted off the top of the fuselage as a unit, and the engine was lowered into position from the top. This proved to be less than ideal in the field, where maintenance necessitated heavy cranes and workstands. (British Aerospace)

the fact that the engine had not gone through a proper development program and was really a "hot rod" variant of the Mk 101. Surge problems at high angles of attack during wing-borne flight quickly surfaced, as did problems during inverted flight. As a consequence, the Mk 103 was further delayed while the engine went through additional testing. Luckily, the new fan section with its transonic blades proved to provide significantly better characteristics than either of the earlier engines, and the Mk 103 was delivered in new-build Harriers from 1971 onwards. Surprisingly, the more powerful engine had a service life rated at 600 hours, much better than the 200 hours of the Pegasus 6. Compared to the original Pegasus 1 (BE.53/2), the Pegasus 11 represented a 139% increase in thrust for a weight increase of less than 60%.

Despite an obvious outward similarity to the Kestrel, the Harrier was virtually a new design structurally, designed for a 3,000 flight hour service life. Yet another new wing was designed for the Harrier, this one being slightly larger with tips that extended past the outrigger fairings. The original Dowty/Rotol landing gear had been designed to absorb 13,000 pounds descending at 8 ft/sec – Harrier would withstand 16,000 pounds at 12 ft/sec. All Kestrels had used the main landing gear door as a speed brake, but the Harrier was designed with a dedicated ventral speed brake. Provisions were made for a retractable in-flight refueling probe on the left air intake, and for the first time the pilots had the benefit of a Martin-Baker Mk 9 ejection seat. The forward facing nose camera of the Kestrels was replaced with an F.95 camera that viewed obliquely through a optical port on the left side of the nose.

Like many of the fighters emerging during the 1960s, the Harrier featured no fixed armament. Provisions were made for mounting two Aden 30-mm gun pods under the fuselage on either side of the centerline weapons stations. The gun pods replaced strakes normally located in those locations to provide a ground cushion during hover. Each 900-pound pod contained a single cannon and 120 rounds of ammunition. The centerline station could carry a 1,000-pound bomb or a reconnaissance pod, and four underwing stations could carry rocket pods, bombs, or fuel tanks. The inboard wing stations were rated at 1,500 pounds, while the outboard stations could carry 750 pounds.

Hawker also developed a set of "ferry" wing tips, although they were seldom seen fitted to aircraft. These fiberglass units could replace the standard wing tips, providing a few square feet of additional wing surface. They were predicted to give a 10% increase in range, but operational use proved to be difficult and the concept was soon abandoned. In practice, the Harrier had to wait for its standard wing tips to arrive at the forward location (usually via a C-130) since the fiberglass ferry tips were not rated for maneuvering flight. And most of the time the Harriers flew in loose formation with a tanker anyway while on ferry flights, so the slight increase in range was a moot point.

The "ferry" wing tips were extensively publicized early in the Harrier's career, and were frequently seen in photographs. They were, however, seldom actually used since they offered no real operational benefit. (British Aerospace)

The Harrier was the first British military aircraft to be equipped with a comprehensive HUD. The system, provided by the Specto Division of Smiths Industries, could display speed, height, and altitude. Other cockpit instrumentation was generally similar to that projected for the abandoned P.1154, and included a Ferranti FE541 inertial navigation system and moving map display. The moving map display was derived from one developed for the cancelled TSR.2 strike aircraft. Maps covering 850 square miles were stored on 35-mm film and displayed on a six-inch diameter display between the pilot's legs. Interestingly, a conventional autopilot was not included. This is because many believed that V/STOL aircraft would require sophisticated stability augmentation systems (SAS), although the Harrier's docile handling ended up requiring only a simple pitch-roll SAS.

The first preproduction Harrier made its initial flight on 31 August 1966 at Dunsfold. All of the preproduction aircraft were used extensively for trials, and the sixth aircraft (XV281) was sent to Sicily for tropical testing during August 1967.

Like all early Harriers, the G.R. Mark 1 was produced in Hawker's Kingston plant. This is a view of the production line during 1969. (British Aerospace)

One of the Harriers was operated from a helicopter platform on the Italian warship *Andrea Doria*, while, ironically, another operated from the Argentinean aircraft carrier *25 de Mayo* as it passed through the English Channel on its delivery voyage from the Netherlands.

The first 27 production Harrier G.R. Mark 1s immediately followed the preproduction aircraft. The first (XV738) made its maiden flight at Dunsfold on 28 December 1967 with Duncan Simpson at the controls. Beginning on 1 April 1969, 20 new Harriers were delivered to No. 1 (Fighter) Squadron at RAF Wittering. On 5 May 1969 two Harriers from No. 1 (Fighter) Squadron flew nonstop from downtown London (near the St. Pancras railway station) to downtown New York (a waterfront site in Manhattan) during the Daily Mail Transatlantic Air Race. Total flight time was 6 hours, 11 minutes, and 57 seconds to cover slightly over 3,000 nm.

The Harrier is capable of operating off of ships other than aircraft carriers. Here a GR1 uses the Royal Navy cruiser HMS Blake *as a landing platform during August 1967.* (Hawker Siddeley Aviation via the Terry Panopalis Collection

No. 1 Squadron continued to open the envelope for the Harrier. Live weapons firing took place at RAF Akrotiri, Cyprus, during the spring of 1970. During 1971 two Harriers were dispatched to the HMS *Hermes* to demonstrate the Harrier's ability to operate from aircraft carriers for a prolonged period. The squadron also dispatched aircraft to Bardufoss, Norway (inside the Arctic Circle), Belize in Central America, and Sardinia.

The RAF originally intended to equip two full squadrons with Harriers – No. 1 (Fighter) Squadron at RAF Wittering in the UK, and No. 4 (Fighter) Squadron at Wildenrath in Germany. In order to provide sufficient pilots, the No. 233 Operational Conversion Unit (OCU) was established at Wittering. As experience was gained with the Harrier, the RAF ordered sufficient quantities to equip two additional squadrons, No. 3 and No. 20, both at Wildenrath, beginning in 1970.

When the Mk 103 engine was retrofitted to the GR1 aircraft, they were redesignated GR1As. During the

Early production Harriers competed in the 1969 Daily Mail Transatlantic Air Race. The goal of the race was to deliver mail from the Post Office Tower in London to the Empire State Building in New York. The Harriers intended to shortcut the normal airports by using a coal yard near St. Pancras railway station and a waterfront helipad in Manhattan. Motorcycles were used for the short intracity distances. A Royal Navy Phantom won the west-bound leg, but a Harrier won the east-bound. (British Aerospace)

The Harrier's claim to fame was its ability to operate from austere airfields. This GR1 shows the minimal equipment Hawker claimed was needed to sustain limited operations – a ladder, toolbox, and a fuel truck (behind the hay-berm). Note the unique inboard pylons for carrying two weapons. (Rolls-Royce Aircraft Engines)

early 1970s, the original twin 4 kva alternators were replaced with a single Lucas 12 kva system that had originally appeared on the US Marine Corps AV-8A. At the same time, a new Lucas 6 kva auxiliary generator was incorporated, but again this change first appeared on USMC aircraft. Various other minor changes improved the reliability and maintainability of the GR1A.

Harrier G.R. Mark 3

By 1973 Ferranti had developed an improved nav-attack system for the RAF's Harriers that incorporated a laser range-finder. The laser ranging and marked-target seeking (LRMTS) system allowed the Harrier to detect targets being designated by other aircraft or ground troops, and also to designate targets itself. A longer nose with a modified profile housed LRMTS, which was similar to the equipment fitted to the SEPECAT Jaguar GR1. The laser was stabilized through a full 360° of roll and ±25°, of pitch. It could be directed as a pencil beam within a 20° cone directly ahead of the aircraft, emitting laser pulses which were adjustable by the pilot between 2 and 4 per second. This allowed different targets to be designated by different aircraft without confusing the receivers. Data from the laser receiver was displayed on the HUD, and also sent to the weapons computer.

At the same time a Marconi ARI.18223 radar warning receiver (RWR) was added to the aircraft to warn the pilot when he was being tracked by hostile radar. The RWR antennas were located in a fairing on on the upper leading edge of the vertical stabilizer and in the tail stinger. The RWR provided both visual and audio indications to the

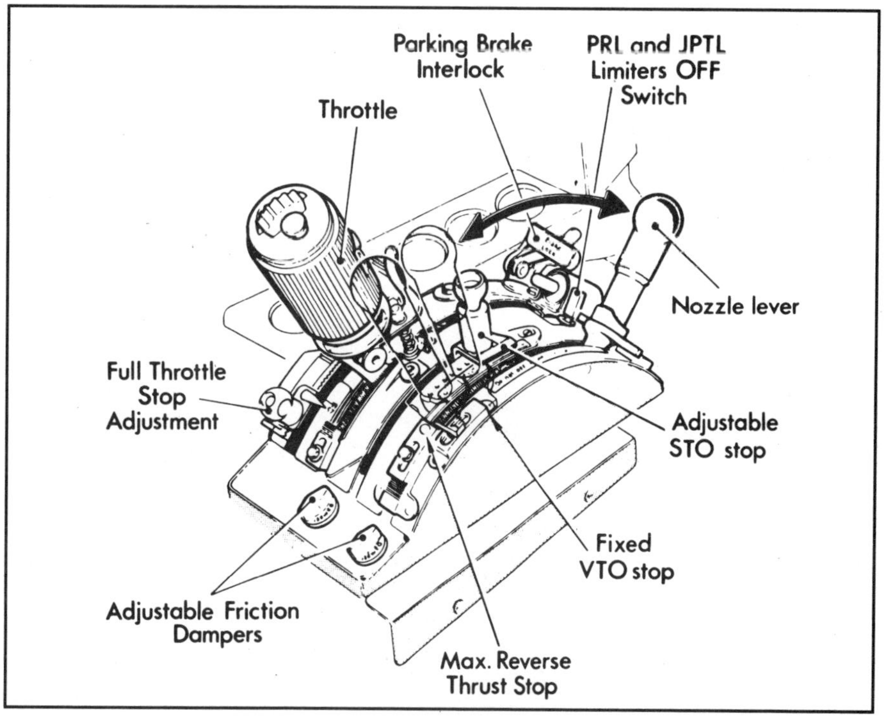

The throttle quadrant for the Pegasus engine has a few features not normally found on jet fighters. The nozzle lever controls the position of all four nozzles, which always operate together to avoid uneven thrust vectors. A pair of stops, one fixed and one adjustable, are provided to limit nozzle rotation. (Rolls-Royce Aircraft Engines)

Twenty years later, XV741 wore small white lettering commemorating its participation in the Daily Mail Transatlantic Air Race. By this time the aircraft had been brought up to GR3 standards with RWR antennas on the leading edge of the vertical stabilizer and in the tail stinger, as well as the extended LRMTS nose. (Kev Darling/BBA)

pilot when he was being tracked. When the RWR antennas were added to the vertical stabilizer, the upper portion of the fin was slightly enlarged to compensate for a slight instability introduced with the longer laser-equipped nose.

The modified aircraft were redesignated GR3, and all remaining GR1 and GR1As were brought up to the standard as they went through depot maintenance.

Contrary to many reports, the Harrier enjoyed a fair reputation as a fighter, at least against its contemporaries. Its high thrust-to-weight ratio when lightly loaded provided a good rate of climb and rapid acceleration. The Pegasus turbofan was fuel efficient, with a maximum fuel flow of only 250 lb/min, compared to 700-800 lb/min for normal 20,000 lbf turbojet fighter engines. At low levels the Harrier could cruise at 450 knots with a fuel flow of around 75 lb/min, enabling it to fly for over an hour on 5,000 pounds of internal fuel.

When the big-wing GR5 Harrier II was finally approved, plans called for equipping the squadrons based

The LRMTS laser is housed behind an optical glass nose, and can gimbel in order to track a target while the aircraft maneuvers. (Ferranti via the Terry Panopalis Collection

The LRMTS nose did not do much for the aesthetics of the GR3, but it added a great amount of capability. Unlike the Americans, the RAF intended to use their Harriers in combat against Warsaw Pact forces and required a much more versatile and capable weapons system. (Nick Challoner)

in Germany with the new aircraft while retaining GR3s for squadrons based in the UK. The GR3s would receive a mid-life upgrade including modifications to extend their fatigue life, a small LERX at the wingroot, enlarged fuel tanks, twin AIM-9 launch rails, a new nav-attack system, updated ECM equipment, and a new anti-armor missile. However, the procurement of 34 additional GR5s signaled the end of the GR3, and the aircraft were retired instead of being modified. This was perhaps just as well since extending the fatigue life had proven harder than anticipated.

Temporarily to Belize ?

Ever since British Honduras became Belize in 1948, it had lived under threat of invasion from Guatemala. When peace talks broke down in October 1975, the RAF dispatched six Harriers from No. 1 Squadron to defend the small country. This was the first overseas deployment for the Harrier and was supposed to last a few months. Except for a brief lapse in 1977, it would last for the better part of 18 years.

Belize was not a friendly location for a V/STOL aircraft, and temperatures in excess of 90°F all but ruled out VTO operations.

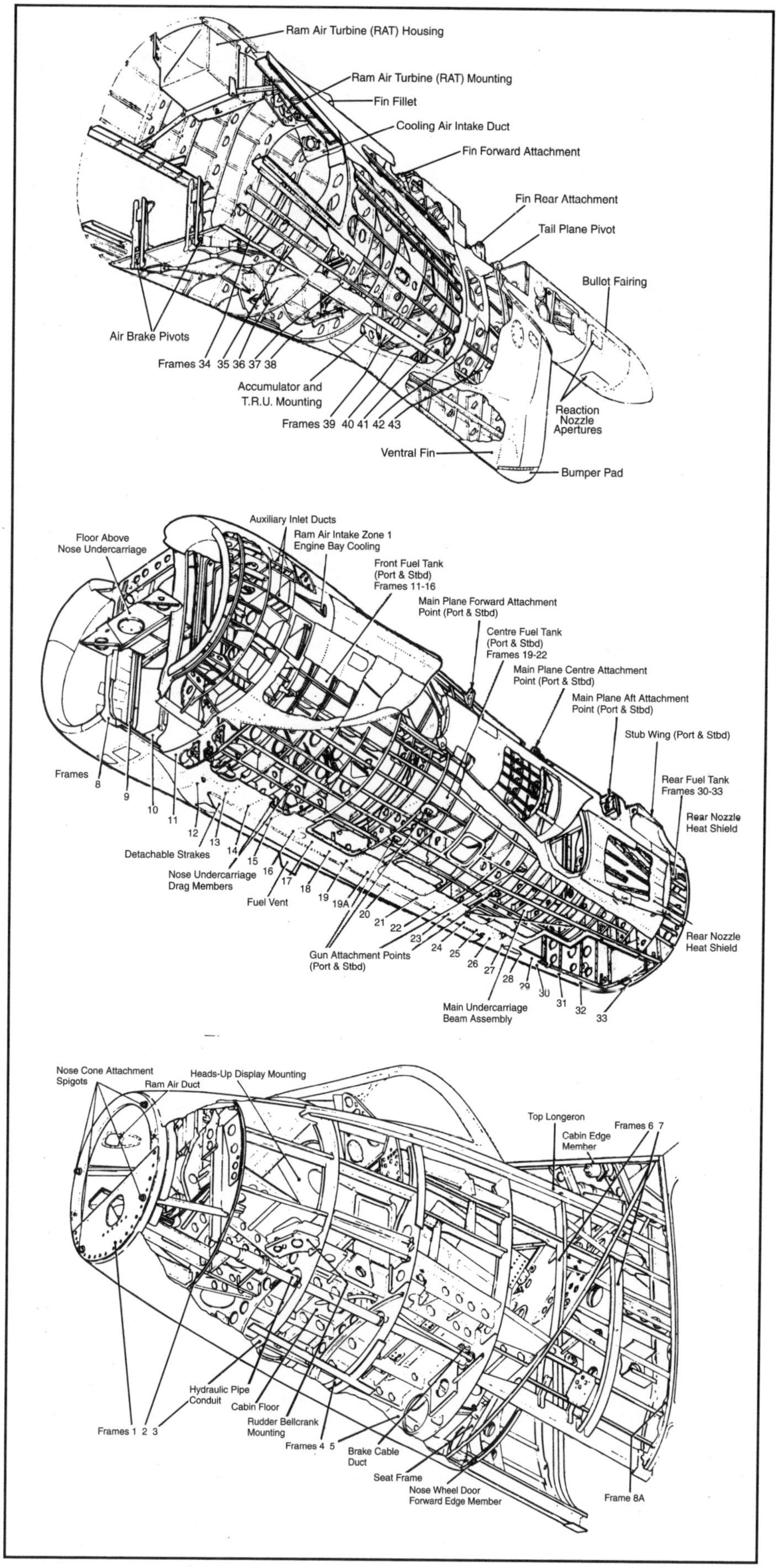

The structure of the Harrier was very conventional, using frames and stringers of aluminum alloy. The basic airframe was designed for a 3,000 hour service life, although various areas were prone to premature stress cracking. This was one of the reasons the RAF decided to retire their GR3s and procure new GR5s instead of undertaking a major retrofit program. (British Aerospace)

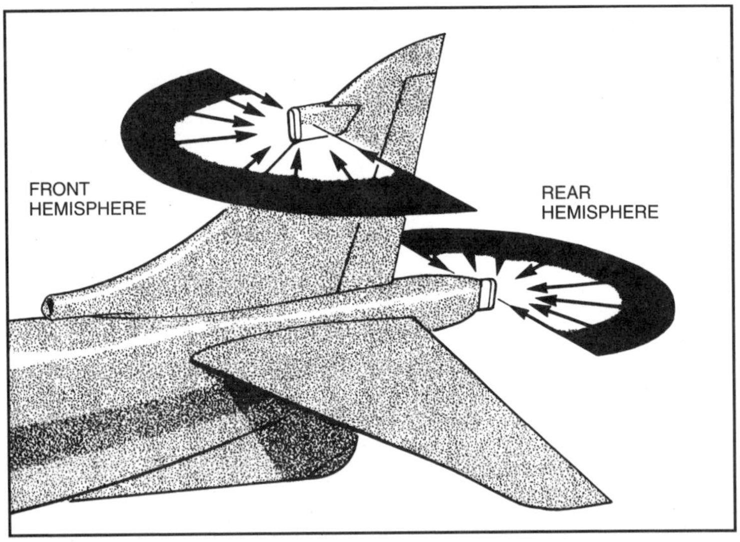

The GR3 upgrade included the addition of a Marconi ARI.18223 radar warning receiver. Two antennas, one on the leading edge of the vertical stabilizer and the other in the extreme rear of the tail stinger provided data to the receiver. Each antenna covered 180° although the forward antennas were somewhat blocked from downward coverage by the wing and fuselage. (British Aerospace)

Harrier GR3 from the No. 1417 Flight of No. 1 (F) Squadron stationed in Belize. (British Aerospace)

Fortunately, the Strike Command Detachment (later renamed the No. 1417 Flight) never had to fight. The RAF rotated pilots and aircraft from all of the Harrier squadrons through Belize at regular intervals, and the remote locations provided excellent training opportunities, including ordnance "live-fire" exercises.

Tensions did finally ease between Belize and Guatemala, and on 8 July 1993 the Harriers left for good, transiting back to Wittering via Key West, Florida. A single GR3 (ZD669) was left as a gate guard at the Belize Airport, just as a reminder of the Harrier's longest temporary assignment.

Despite the length of the "temporary" deployment, the conditions in Belize remained quite primitive. This GR3 undergoes maintenance under a camouflage net. Note the "palm tree" on the vertical stabilizer. (MoD via the Terry Panopalis Collection)

The SNEB rocket was an early favorite of the RAF, and was used very effectively during the Falklands. The unguided rocket was largely withdrawn from RAF service by the late 1980s, although the No. 1417 Flight in Belize continued to use the weapon until the Harriers were withdrawn in July 1993. (British Aerospace)

The No. 1417 Flight from No. 1 Squadron operated Harrier GR3s from the Belize City Airport from November 1977 until July 1993. The Harriers were there to protect Belize (the former British Honduras) from Guatemala. Interestingly, a deployment to Belize in November 1975 had marked the first operational deployment of the Harrier GR1, and the withdrawal from Belize marked the last operational GR3 usage. (Royal Air Force)

When the LRMTS nose was added, the pitot tube had to be moved to a location immediately above the laser nose, although the F.95 camera remained in the same location on the left side of the nose. The vertical stabilizer on the GR3 was slightly higher than on the GR1 to compensate for a minor instability caused by adding the LRMTS nose. Interestingly, the vertical would have had to be extended even further if the RWR antennas had not been added at the same time. (Nick Challoner)

When the decision was finally made to retire the remaining GR3s in favor of procuring additional GR5/7s, the end came quickly for the older Harriers. Here is the remaining hulk of XV740 which had served with No. 1 (F) Squadron. Remains of several Canberra bombers can also be seen in the background. (Kev Darling/BBA)

This GR3 (XV759), which had served with the 233rd OCU at RAF Wittering, is heading to St. Athan for use as a fire trainer. (Kev Darling/BBA)

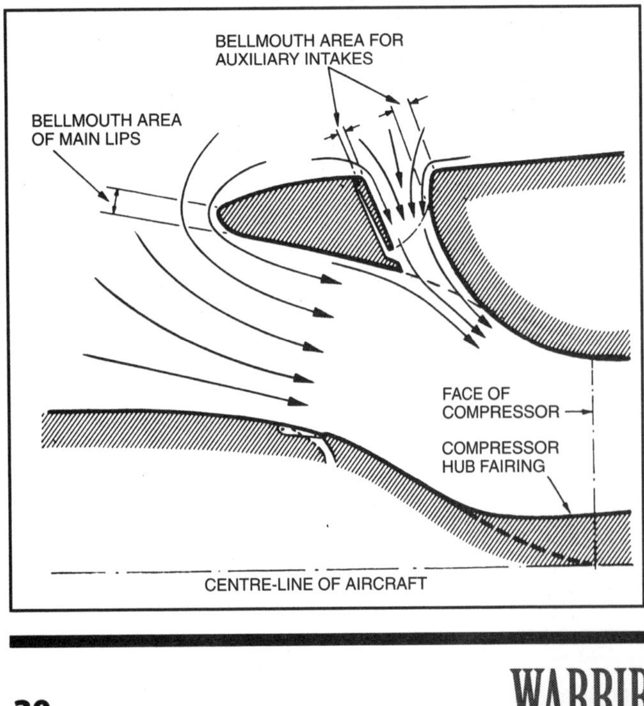

The Pegasus engine in the GR3 consumed over 20,000 pounds of air per minute during hover. The challenge facing Hawker was to design an intake that could supply this amount of air during hover, yet not generate too much excess drag at high speed. The solution was a set of pressure-relief doors on the side of the intake. The doors do not have actuators, and are simply "sucked in" when the pressure inside the inlet is less than the pressure outside, such as is the case during low-speed flight and hover. As it turned out, the total efficiency of the inlet is improved at all airspeeds since the doors operate completely automatically. Each of the doors operates independently, explaining why photos of the Harrier show different configurations open. The Harrier inlets underwent an extensive development and test series, and the final result was over 97% efficient at most airspeeds. (British Aerospace)

5 Dress Blues

Harriers for the US Marine Corps

The accepted mantra of fighter forces worldwide during the late 1960s and early 1970s was "bigger, faster, heavier," so it is surprising that the RAF supported the Harrier at all. The GR1 was still a subsonic aircraft with a limited range and weapons carrying ability. However, their versatility allowed them to operate from austere sites after, it was reckoned, the RAF's main runways had been destroyed. From these austere sites the Harrier could provide limited close air support to ground troops while the main airfields were rebuilt.

It was just this capability that impressed the US Marine Corps. The Marines had always wanted a fixed wing aircraft capable of deploying with the troops to the battle front, and not suffering the transit time of carrier-borne aircraft based offshore. In a rare instance of corporate marketing actually paying off, a promotional film was shown to the US Marine Corps' Deputy Chief of Staff (Air), MajGen. Keith McCutcheon during late 1967. McCutcheon was impressed enough to send two pilots, Col. Tom Miller and LtCol. Bud Baker, to observe the Harrier at the 1968 SBAC Show at Farnborough, and within two weeks the pair had arranged to fly Harriers at Dunsfold.

The two pilots had several complaints about the Harrier – brake chatter during taxi, tailplane vibration at low speed, the need to remove the wing in order to service the engine, and generally poor rearward visibility. But overall they were impressed, and they returned to Washington full of enthusiasm for the new aircraft. Uncharacteristically for Washington, things happened quickly and a US Navy team was performing a Navy Preliminary Evaluation (NPE – the first step to the Navy accepting an operational aircraft) within three months. After further evaluations, the US Navy ordered 12 AV-8A Harriers (BuNo 158384/158395) for the Marines.

Hawker called the USMC Harrier the Mk 50, and the initial 12 aircraft were essentially identical to the GR1 being produced for the RAF. This included the Pegasus 102 engine and Ferranti FE541 inertial nav-attack system. This and further orders from the Marines would total 102 AV-8As and 8 two-seat TAV-8As (Mk 54).

This VMA-231 AV-8A (BuNo 159243) is representative of the appearance of all Marine Harriers in late 1975. As part of the "off-the-shelf" procurement strategy, even the British camouflage was kept. Noteworthy are the large full-color stars-and-bars insignia on the air intake and wings. The VHF/FM radio antenna on top of the wing was one of the few external differences between the GR1 and AV-8A. (Mick Roth Collection)

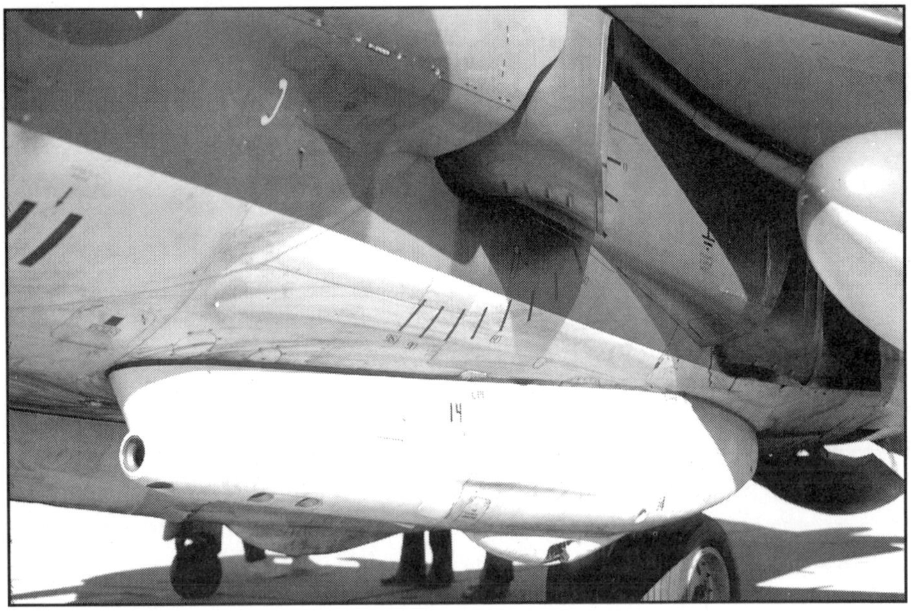

AV-8As carried the same Aden 30-mm cannon as the RAF – creating a logistics nightmare for the Marines since the ammunition was not used by any other US weapons system. The cannon were mounted in two pods that replaced the normal fuselage strakes. The strakes, and cannon pods, also serve as aerodynamic devices to trap ground effect air, thereby increasing lift during hover. (Mick Roth)

It is very seldom that a non-US aircraft is approved for service with the US military. Congress was not pleased that significant defense dollars would be spent on a "foreign" aircraft, and subsequently Hawker Siddeley approved a 15-year license agreement with McDonnell Douglas for the production of Harriers in the US. This agreement included any future derivatives of the basic Harrier configuration developed by either country. But during the FY71 budget negotiations, Congress balked at paying the startup costs associated with transferring Harrier production to McDonnell Douglas. Subsequent production years ran into the same problems, and in the end all AV-8s were manufactured by Hawker Siddeley at Kingston. After manufacture they were test flown by Hawker Siddeley pilots, then partially disassembled and transported by C-141 or C-5 to St. Louis where they were reassembled and again test flown prior to delivery, this time by McDonnell Douglas pilots.

Partially to get around the Congressional objections, the Harrier was purchased as an "off-the-shelf" system. Very few changes were incorporated into initial aircraft, mainly consisting of the incorporation of a weight-on-wheels sensor (common to all American military aircraft), slightly different checklist placards in the cockpit, and the substitution of in-service Navy avionics (VHF/FM, UHF, and IFF) in lieu of the RAF equipment. The new radios were usually associated with a large tac-

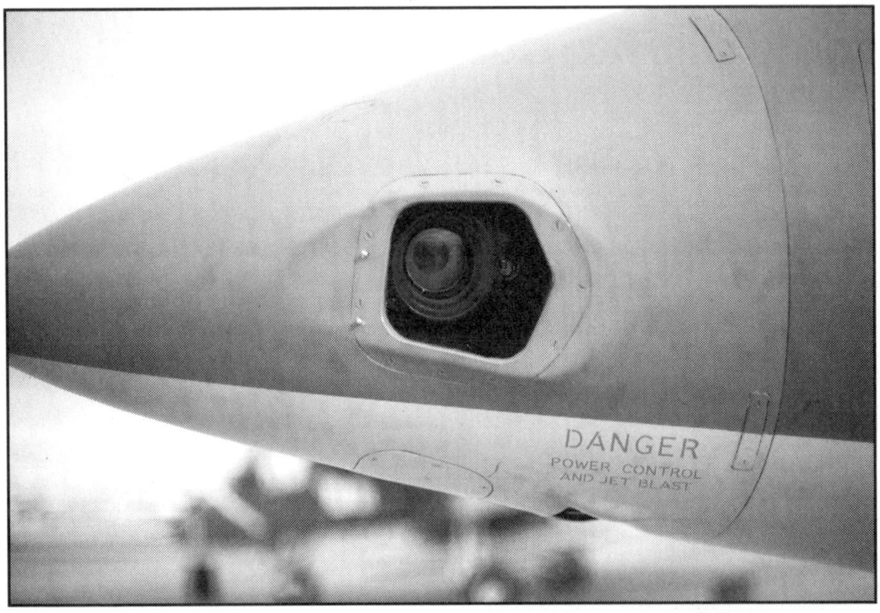

The AV-8A carried the same F.95 camera used by the RAF, and the Marines found it to be equally as useless for battle damage assessment. It did provide a limited reconnaissance capability for the Harrier, but the photo quality was not particularly good. The lower reaction control valve can be seen below and slightly behind the camera port, along with a red warning painted on the side of the fuselage, important since ground crews were not used to having compressed air exhausted near the nose of an aircraft. (Mick Roth)

One change from the RAF Harrier was the ability to carry AIM-9 Sidewinder missiles. The change involved adding wiring for the missile to the outboard pylons. The particular way Hawker accomplished this made the pylon non-removable, so AV-8As are always observed with the outer pylon in place, unlike the GR1. (Mick Roth Collection)

tical VHF antenna above the wings on USMC aircraft. The first 12 aircraft were powered by Pegasus 102 engines (designated F402-RR-400 by the US), but all other AV-8As (and TAV-8As) were powered by the improved 21,500 lbf Pegasus 103 (F402-RR-401) found on the later RAF G.R. Mark 1A.

One other change was made at the Marines' request. The outer pylons were equipped with wiring that enabled them to carry the AIM-9 Sidewinder air-to-air missile, in addition to the rockets, bombs, etc. that the RAF version could carry. This made the outer pylon non-removable, unlike the pylons on RAF aircraft. The Marine Corps had given up an opportunity to purchase 17 additional McDonnell Douglas F-4J Phantom IIs in order to procure the initial dozen Harriers, and they did not want it to appear they had bought an aircraft incapable of defending itself. As part of the purchase agreement,

The Harrier came equipped with a ventral speedbrake, unlike the earlier Kestrel which had used its main landing gear door as a speedbrake. (Mick Roth)

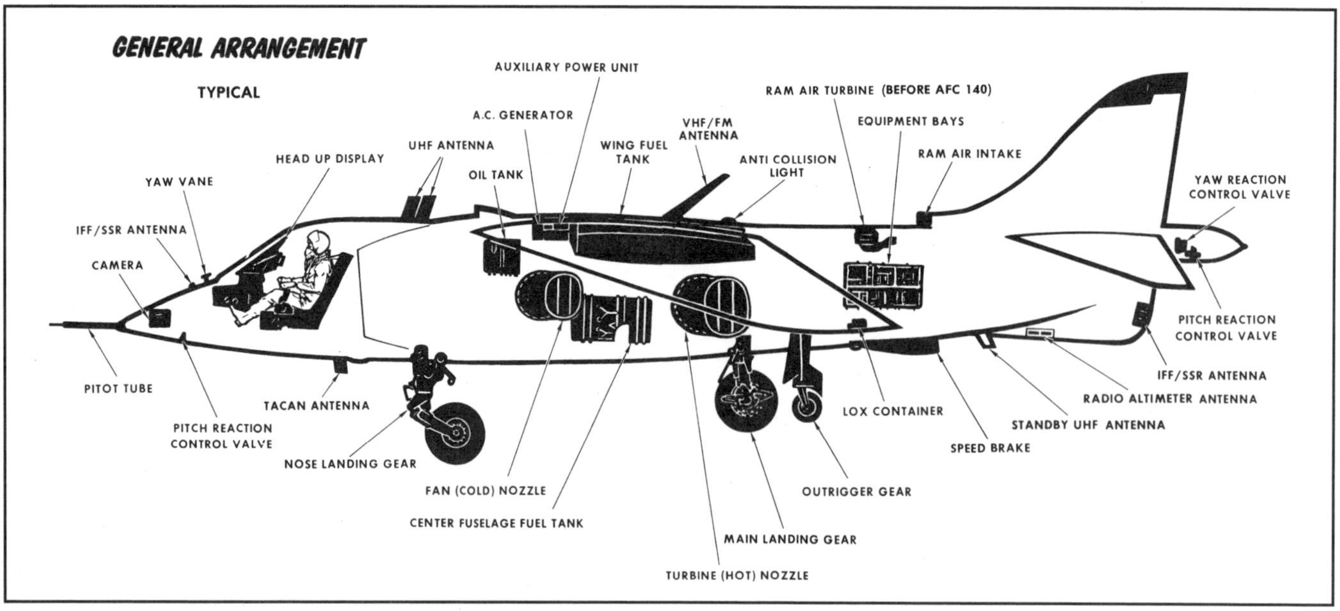

The general arrangement of the AV-8A was also representative of the GR1 since they were essentially identical. The ram air turbine (RAT), located on the upper fuselage ahead of the vertical stabilizer, was removed later in the AV-8s career since it offered little real capability and required a great deal of maintenance. The location of the reaction control valves is also illustrated. (U.S. Marine Corps)

the RAF agreed to fund weapons separation trials for a wide range of American ordnance, something that probably would have been accomplished at some point anyway since the RAF and RN frequently uses US and NATO weapons.

In early 1971 a US Navy Board of Inspection and Survey Trials was conducted at NAS Patuxent River, clearing the Harrier for operational use. The first batch of aircraft was assigned to VMA-513 at MCAS Yuma, Arizona, which was commanded by then-new Col. Bud Baker. Shortly thereafter, approval was granted to operate the Harrier from LPH and LPD (amphibious assault) ships, the first jet-powered, fixed-wing aircraft to do so. Eventually AV-8As also equipped VMA-231 and VMA-542 at MCAS Cherry Point, North Carolina.

The pressure relief (also known as "suck-in") doors around the air intake were one of the keys to the Harrier's success as a V/STOL aircraft. These doors provided extra airflow during hover without creating extra drag during high-speed flight. The Pegasus "cold" nozzle under the wing leading edge is fully horizontal. Markings on the fuselage provide reference for the nozzle's angle. (Mick Roth)

Like the GR1, the AV-8A suffered from too little air conditioning capacity, making it an uncomfortable aircraft to fly in warmer climates. The Sea Harrier was the first Harrier to overcome this, simply because the Hawker designers overcompensated when they added environmental capacity for the radar systems. (Don Logan via the Mick Roth Collection)

SERVICING DIAGRAM

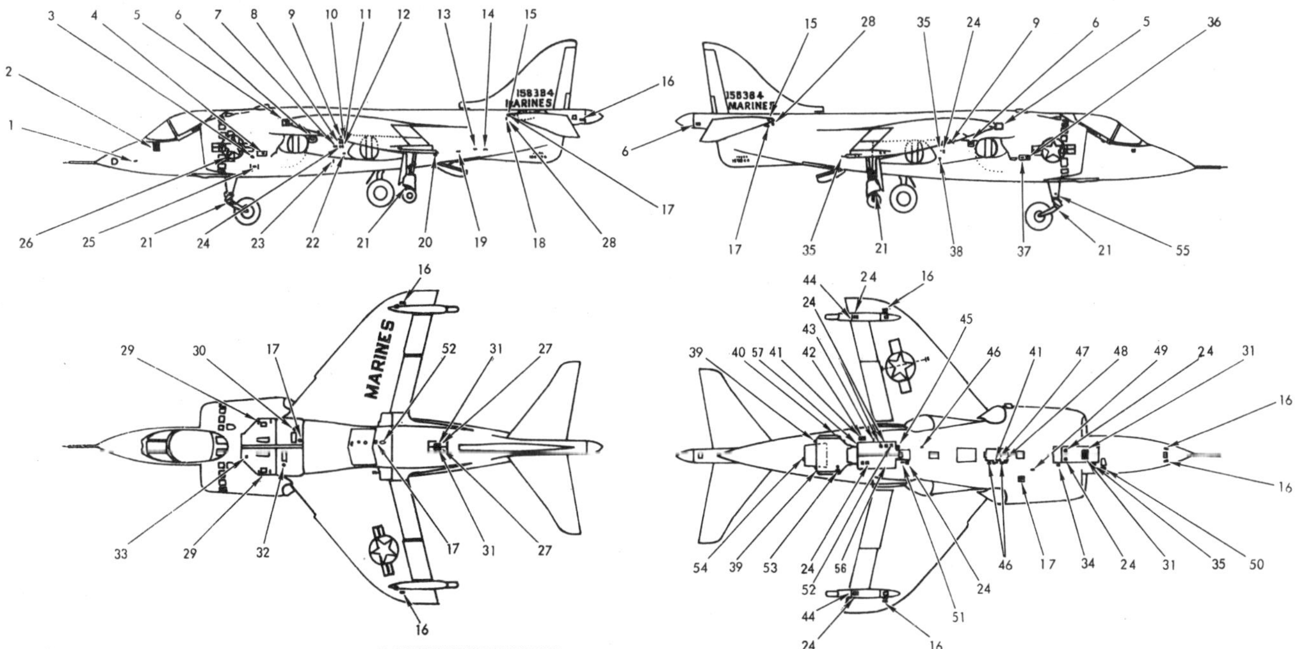

1 WINDSCREEN WASHING FLUID DOOR 7
2 NORMAL CANOPY RELEASE
3 INTERCOMMUNICATION DOOR 20
4 NO. 1 HYDRAULIC SYSTEM DOOR 22L
5 FIRE ACCESS
6 HYDRAULIC RESERVOIR CONTENTS SIGHT DOOR 34L/R
7 REFUELING AND DEFUELING PRESSURES
8 GROUND (EARTH) HERE
9 PRESS TO ILLUMINATE HYDRAULIC RESERVOIR
10 FUEL DOOR 37L
11 RESERVOIR BLEED VALVE DOOR 108L/R
12 ENGINE OIL CONTENTS CHECK DOOR 37L
13 A.P.U. OUTPUT (NOT USED)
14 SERVICING GROUND SUPPLY 200 VOLTS 400 CYCLES DOOR 57
15 TO AVOID STABILATOR DAMAGE ENSURE FASTENERS FLUSH AND TORQUE LOADED TO 45 LB IN DOOR 61L/R
16 POWER CONTROL AND JET BLAST
17 HYDRAULIC FILTER DOOR 61L/R AND 27L AND 47 AND 24
18 STABILATOR ANGLE MARKINGS
19 AIRBRAKE GROUND LOCK
20 LIQUID OXYGEN 5 LITERS (BEFORE AFC 85)
21 TIE DOWN POINT
22 NO. 1 HYDRAULIC RESERVOIR PRESSURE DOOR 108L
23 DEFUELING SHUTOFF
24 NITROGEN
25 ENGINE OIL DOOR 21L
26 HYDRAULIC HAND PUMP DOOR 16L
27 R.A.T. ACCESS REMOVE 4 BOLTS (BEFORE AFC 140)
28 PANEL MUST BE FITTED BEFORE MOVING STABILATOR DOOR 61L/R
29 FIRE ACCESS
30 GAS TURBINE STARTER/APU INLET
31 KEEP CLEAR
32 GAS TURBINE STARTER/APU EXHAUST
33 CABIN AIR FILTER DOOR 17L
34 TIRE PRESSURE 90 P.S.I. (6.33Kg/cm^2)
35 PITOT DRAIN
36 NO. 2 HYDRAULIC SYSTEM DOOR 22R
37 DEFUELING PRESSURE CONNECTION DOOR 22R
38 NO. 2 HYDRAULIC RESERVOIR PRESSURE DOOR 108R
39 ENGAGE AIRBRAKE GROUND LOCK BEFORE REMOVING DOOR 56 OR ENTERING BAY
40 OXYGEN VENT (STAND CLEAR WHEN CHARGING)
41 FILTER AMSU DOOR 35
42 EQUIPMENT BAY COOLING
43 ARMAMENT SAFETY KEY
44 TIRE PRESSURE
45 WATER DRAIN
46 FUEL TANK WATER DRAIN
47 AIR MOTOR DOOR 35
48 AIR MOTOR FEED PIPE INDICATOR
49 ENGINE FUEL DRAIN TANK
50 NOSE LANDING GEAR DOORS MANUAL OPENING VALVE DOOR 12
51 TIRE PRESSURE 90 P.S.I. (6.33Kg/cm^2)
52 WATER
53 LOX CONTENTS GAUGE
54 BATTERY DOOR 56
55 HYDRAULIC FILTER
56 EXTERNAL POWER RECEPTACLE (L. SIDE MWW)
57 LIQUID OXYGEN 5 LITERS (AFTER AFC 85)

SPECIFICATIONS

		NAVY	NATO
FUEL	PRIMARY	MIL-T-5624E, JP-5	F-44
	ALTERNATE	MIL-T-5624E, JP-4	F-40
	ALTERNATE	NONE	F-34
	ALTERNATE	NONE	F-35
OIL	ENGINE	MIL-L-23699 A and B	0-156
OIL	CSD	STERNOL 1842B	
HYDRAULIC FLUID	LIQUID	MIL-H-83282A	H-515
NITROGEN	GASEOUS	BB-N-411 TYPE I CLASS 1 GRADE B	
OXYGEN	LIQUID	MIL-O-27210 TYPE II	
WINDSHIELD AND CAMERA WASHER FLUID	LIQUID	KILFROST WWF/2	

The servicing diagram shows how to access most of the major systems aboard the AV-8A. (U.S. Marine Corps)

Like the RAF, the Marines quickly fitted larger external fuel tanks to the Harrier to allow it to ferry further. One of the Harrier's major limitations was its relatively short range on internal fuel. Also like the RAF, the Marines purchased the bolt-on refueling probe that allowed the AV-8A to take on fuel via the Navy-standard probe-and-drogue system. Unless it was needed for a particular mission, the probe was seldom carried. (Don Logan via the Mick Roth Collection)

After a year or so of operational service, the Marine Corps began to modify their aircraft. The Ferranti FE541 nav-attack system had never been particularly well liked by the Marines, but it had been purchased as part of the "off-the-shelf" philosophy. The Marines felt the system was too complex and took too long to set up before flight for the relatively simple attack missions they intended to

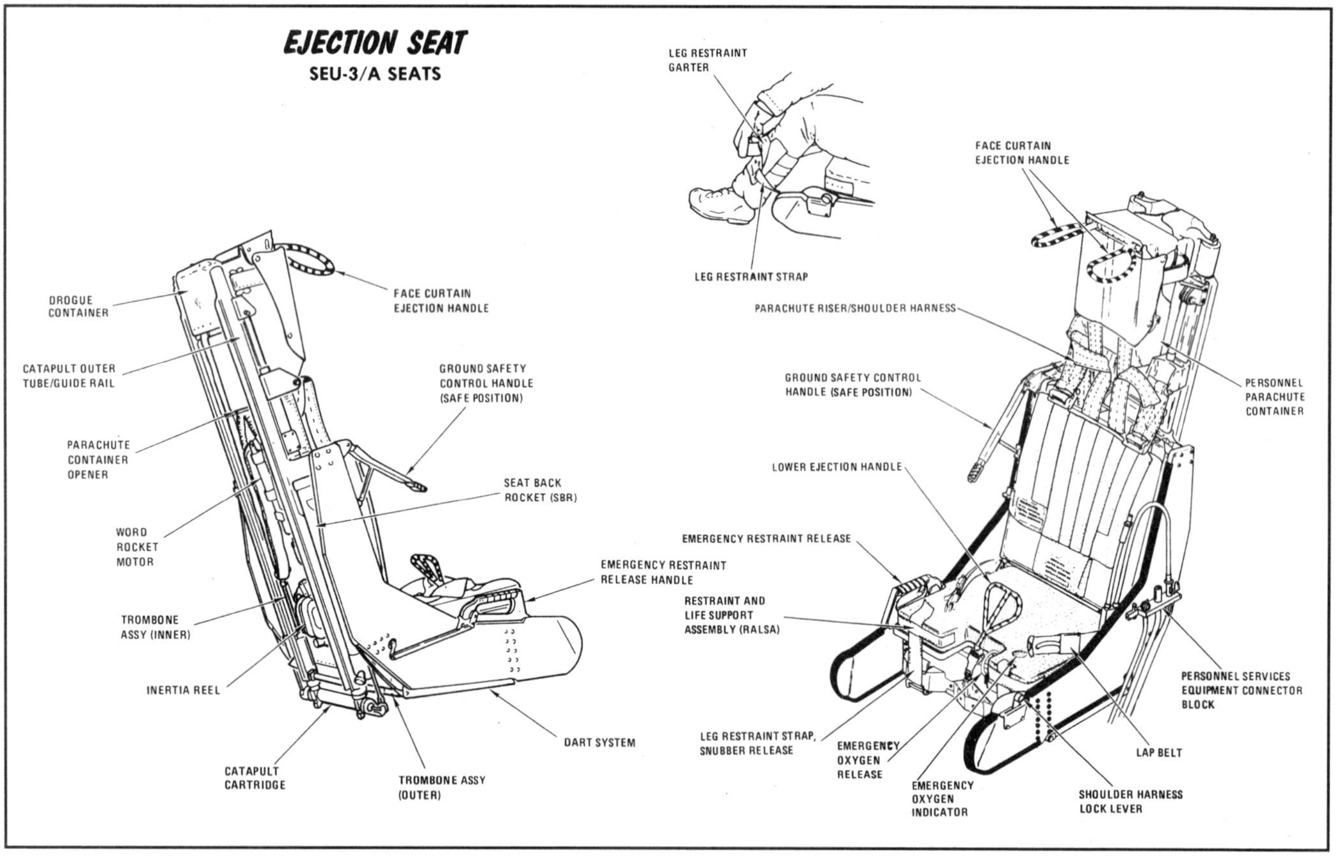

One of the changes specified by the Marines was the use of a Stencel SEU-3/A (SIIIS-3) ejection seat. This was not due to any perceived flaw in the Martin-Baker Mk 9A seat used by the RAF, but rather an attempt by the US Navy to standardize one seat across all of its fighter aircraft. As it turned out, sufficient funds were never found to install the Stencel seat in the F-14 and F/A-18, leaving the AV-8A as the sole user. (U.S. Marine Corps)

An AV-8A during transition from vertical to wingborne flight. Note the position of the forward nozzles. (Mick Roth)

Energy Levels in V/STOL Flight

The energy output from the four nozzles in V/STOL flight is about 30,000 horsepower. The reaction controls at full control demand have an energy output of several thousand horsepower. The front nozzles exhaust emerges at about 700 knots, 105°C (220°F), and 16 psi. The rear nozzles exhaust emerges at about 1,050 knots, 645°C (1,195°F), and 11 psi. The reaction control valves exhaust emerges at about 1,500 knots, 400°C (750°F) and 150 psi. Although velocity, pressure, and temperature drop off with distance, the exhaust velocity at ground level in a low hover can be 300-400 knots at 4 psi. If this pressure is permitted to build up under a surface such as a landing mat or manhole cover, the lifting force becomes tremendous. A pressure of 4 psi will lift 4-foot-thick concrete or 8-inch-thick steel. The Harrier has proven to be an efficient manhole cover remover, although it displays no discretion on depositing them after removal. The aircraft has raised an 11-ton mat 4 feet above the ground. Pneumatically supported mats do not soften the landing; therefore, all landing mats should be thoroughly sealed including the perimeter. The aircraft should never cross the edge of a mat in V/STOL flight at less than 50 feet. (Excerpt from AV-8A Flight Manual)

VMA-542 AV-8As operated off HMS Hermes *during January 1978.* (Royal Navy via the Terry Panopalis Collection)

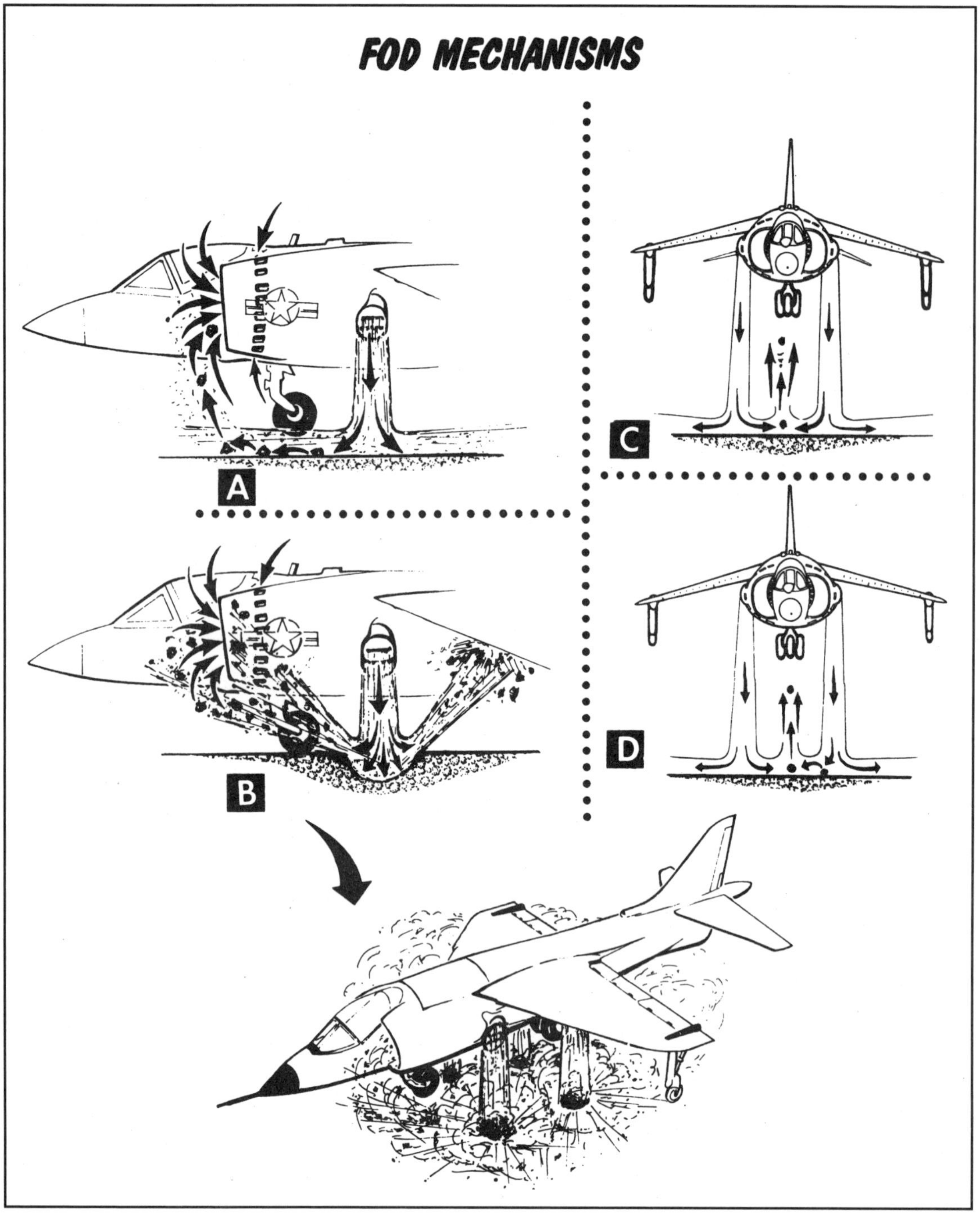

V/STOL aircraft are particularly adept at creating their own Foreign Object Damage (FOD) and then ingesting it. The above figure illustrates four mechanisms for FOD ingestion. (A) Even a smooth surface is relatively bumpy to a small object such as a rock, bolt, gravel,, etc. As this object is rapidly accelerated by the radial flow of perhaps 400 knots of the jet exhaust, it soon encounters a bump and bounces up, often into the engine intake flow which is also at a high velocity, and thus into the engine. (B) On an unprepared surface, the Harrier is capable of rapidly digging its own grave. As a crater is formed, the crater itself serves to direct the ejected material back towards the aircraft. Although the illustration shows the aircraft on the ground, the 8 to 10 psi of the jet exhaust in a low hover will immediately form a crater in the loose soil. On a grass surface, the jet temperature will dry the soil in a few seconds and shrinkage cracks will normally appear. Clods and chunks are then blasted out of the soil. For this reason, truly vertical VTOL should never be made on an unprepared surface. (C) If an object lies at the stagnation point between two opposing radial jet sheets, it will be subjected to a large negative pressure gradient which will give it an upward acceleration of approximately 3-g and will enter the jet foundation. Thus, the interaction of the jet exhausts form a giant vacuum cleaner effect. (D) An object may be bounced into the jet foundation in an action similar to that in A and then will be accelerated upward as in C. (Excerpt from AV-8A Flight Manual)

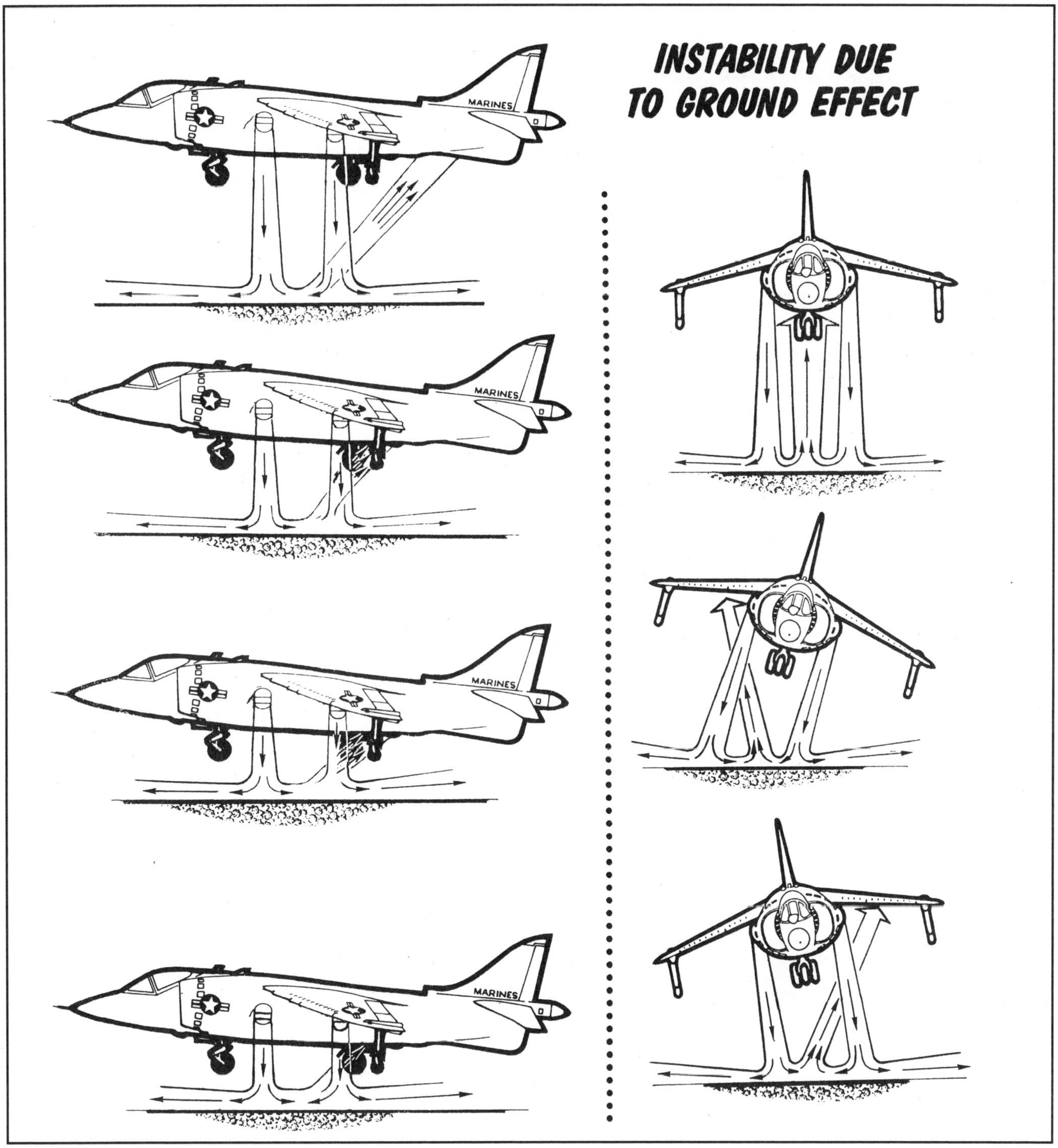

Two instability mechanisms are associated with the jet fountain. As the aircraft reaches a critical altitude the jet fountain moves forward from aft of the aircraft and commences to impinge on the tail surfaces, causing a nose down trim change. As the aircraft descends, the center of pressure moves forward on the aircraft and at the same time, becomes more powerful. These two actions tend to cancel each other, however, as the jet fountain moves forward, its force is expended on varying surface areas resulting in random pitch trim changes. As the aircraft is rolled, the jet fountain moves towards the high wing. If it then impinges upon the aircraft it will tend to increase the roll angle and may cause a pitch trim change. If it leaves the aircraft surface it will cause a nose up pitch trim change. Surface irregularities will also cause deflection of the jet fountain causing rapid trim changes or turbulence sometimes known as cobblestoning. Wind gusts will also cause random trim changes. (Excerpt from AV-8A Flight Manual)

fly. All of the FE541 components were eliminated beginning with the 60th aircraft, although the Smiths HUD was retained. A new US-developed two-gyro attitude and heading reference platform was fitted, and Smiths produced a modified interface/weapon aiming computer (I/WAC) which solved the ballistic equations and generated the appropriate symbology for the HUD.

This was known as the "Baseline Attack System" and provided very limited weapon aiming modes, but it met the modest mission requirements of the Marines more aptly than the sophisticated FE541 which had been designed for all-out war in Europe. At the same time the cockpit layout of the AV-8 was revised to be more in keeping with US norms. A more sophisticated attitude reference platform was later introduced, and the I/WAC was modified to incorporate additional weapons aiming modes.

A second change was replacing the Martin Baker Mk 9A ejection seat with a Stencel SEU-3/A seat. This was not because of any particular shortcoming of the British seat, but was driven by an effort to standardize ejection seats across the Navy. The new seat was installed beginning with the 90th AV-8A, and by mid-1977 had been retrofitted to all earlier aircraft. A plan to install a similar seat in F-14 and F/A-18 aircraft never materialized.

Other changes resulting from operational experience included additional hydraulic filters and the removal of the ram air turbine. The Marines spent a larger percentage of their time operating from unprepared areas than the RAF, and hydraulic contamination had been a problem. Hawker Siddeley accomplished the engineering, and these changes were incorporated into the last AV-8A while it was on the production line, and earlier aircraft as they passed through depot-level maintenance.

In an effort to get as many single-seat Harriers as possible before Congress decided to stop funding purchases of a foreign aircraft, the Marine Corps waited until their last

A 1977 photograph of Harriers aboard an aircraft carrier. Although a relatively small aircraft, the Harrier nevertheless presented some handling difficulties aboard large-deck carriers. One example is that since the outrigger gear were near the wingtips, it was difficult to place a wing over the deck edge. (Geoff Rhodes via the Mick Roth Collection)

MINIMUM TURNING RADIUS AND GROUND CLEARANCE

NOTES

TURN USING MAXIMUM NOSE WHEEL STEERING ANGLE (45°)

OFFSET DISTANCES (180° TURN):
℄ TO ℄ 32 FEET.
OUTBOARD OUTRIGGER 54 FEET 3 INCHES
OUTBOARD WINGTIP 57 FEET 3 INCHES.

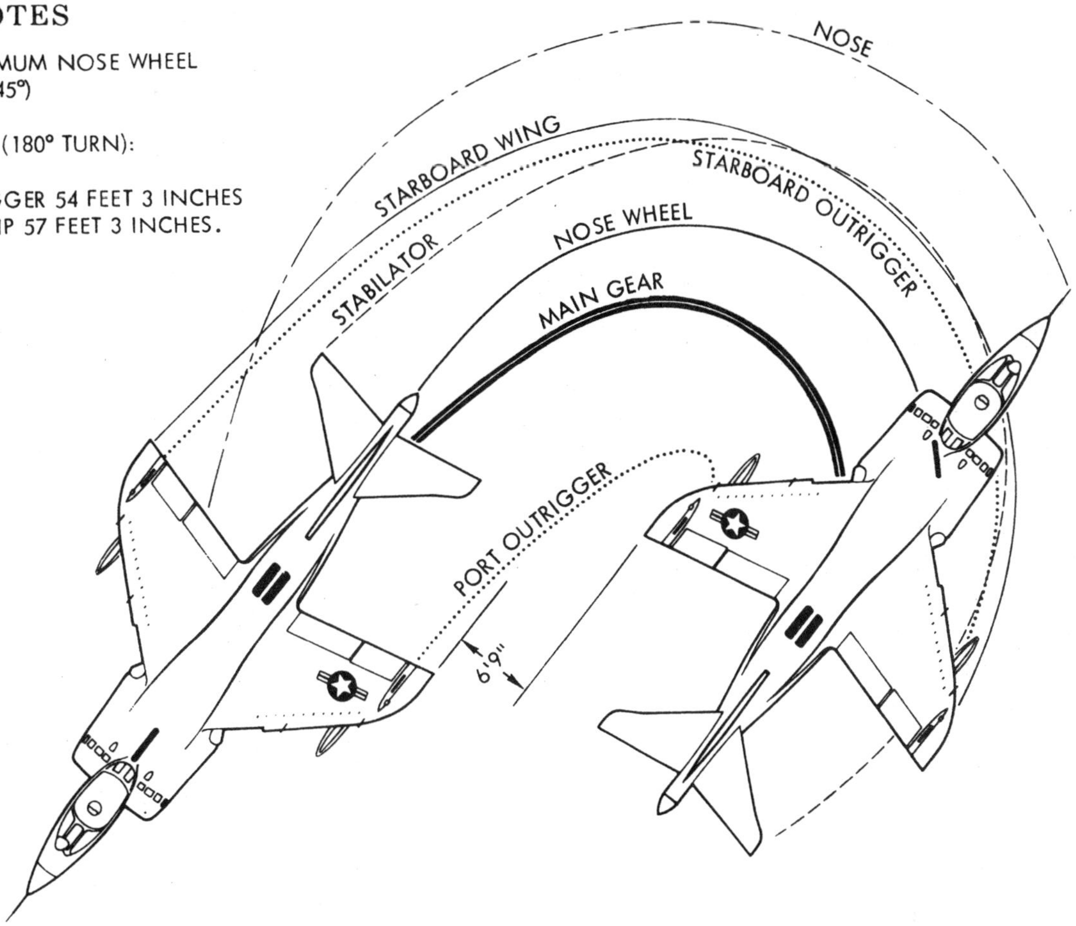

TAV–8A

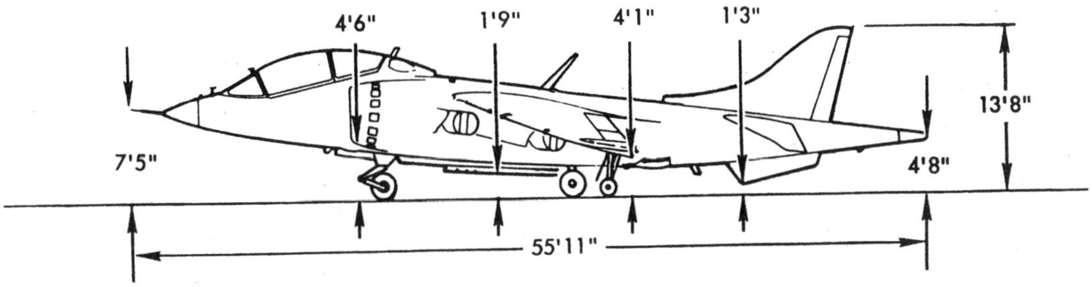

AV–8A

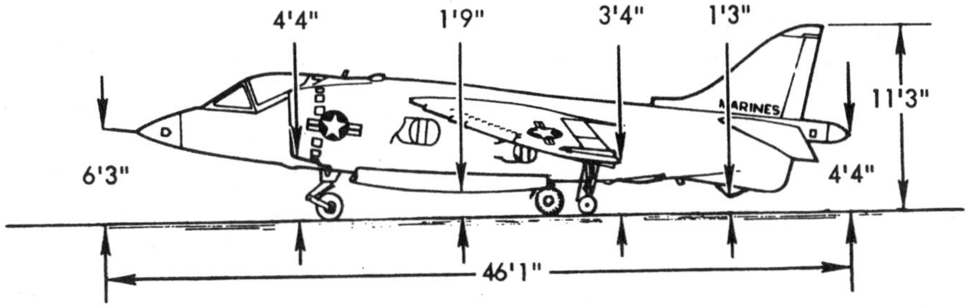

One AV-8C was transferred to NASA after it was retired from Marine service. Although assigned to the Langley Research Center, it was photographed at Edwards AFB on 6 October 1990. Other than the gloss white and blue paint scheme, little had been changed from its Marine days. (Craig Kaston via the Mick Roth Collection)

Sometimes the Harriers would have special camouflages applied with water-soluble paint for specific exercises or deployments. This AV-8C (BuNo 159247) from VMA-542 wears a brown and tan paint scheme while at Nellis AFB on 17 November 1984. (Mick Roth Collection)

order to procure any two-seat trainers. When the TAV-8A was finally ordered, it incorporated all the changes previously made to Marine Harriers, including the revised Baseline Attack System, Stencel ejection seats, and the various mechanical improvements. In addition, the TAV-8A was to operate as a "tactical air command" aircraft, and provisions were made to fit additional VHF and UHF radios in the rear cockpit. The first TAV-8As were delivered to VMA(T)-203 at Cherry Point during the summer of 1976.

An early AV-8A from VMA-513 demonstrates its ability to hover. (Terry Panopalis Collection)

The Marines were serious about using the Harrier as an air defense fighter, even though it lacked any meaningful fire control system. Using the aircraft in ways never envisioned by the Hawker designers, the Marine pilots developed tactics around rotating the nozzles and using the reaction control system in wing-borne flight. Known as "VIFFing" (vectoring in forward flight), the Harrier proved remarkably agile in the air combat arena.

The Marines took VIFFing so seriously that the concept was even discussed in both the flight manual and the tactical manual. The manuals addressed the concept in some detail and reminded pilots that although VIFFing could be used to effectively increase controllability at higher than normal angles of attack, extra care must be exercised to fully understand the ramifications of using thrust instead of wing lift during these maneuvers. It also noted that under certain circumstances it was very easy for the aircraft to depart from controlled flight. Recovery involved neutralizing the controls, setting the nozzles to aft, and the throttle to idle.

After the AV-8A/Cs were retired in 1987, several were preserved in museums and as gate guards. This AV-8C (BuNo 158976) is preserved at MCAS Cherry Point. (Craig Kaston via the Mick Roth Collection)

Although in the end no one doubted that the AV-8 was a useful combat aircraft, and it had certainly won the affections of its pilots, the Harrier was nevertheless a rather limited performer. The AV-8A could carry 1,500 pounds of bombs on a 435-mile combat mission using a 1,200-foot takeoff run. This was about 80 miles short of the A-4's normal radius of action, and the A-4 could carry 4,000 pounds of bombs – more than double the Harrier's load. But, the Skyhawk required a traditional runway (or aircraft carrier), and the Marines were happy to trade a limited load for the ability to use the V/STOL capabilities of

Harrier. In actuality, in any given amount of time the Harrier could deliver as much ordnance to most targets as the Skyhawk. It required more sorties to do so, but since it could be stationed much closer to the front, each sortie took a great deal less time.

Toward the end of their useful lives, several modifications were made to USMC Harriers under technical directive AFC-181 to ensure their continued effectiveness until the new AV-8B was delivered. A Marine report titled "Operation Frosty Nozzle" had pointed out several shortcomings with the AV-8A that needed to be addressed immediately. Among these were the lack of electronic countermeasures equipment and secure voice radios. Apparently DoD management agreed, and a CILOP (conversion in lieu of procurement) program was authorized for all remaining AV-8As.

Between 1979 and 1984 a total of only 47 AV-8As were modified with ALR-45F/APR-43 (essentially a poor-man's ALR-67) radar warning receivers, ALE-39 expendable countermeasures dispenser in the rear fuselage, provisions for an ALE-37 chaff dispenser pod on the centerline station, I, and on-board oxygen generation system (OBOGS).The Americans did not like the positioning of the GR3 RWR antennas, and opted instead to locate them in the tail stinger, wingtips, and nose. Most of the lift improvement devices (LID) developed during the AV-8B program were added to the aircraft to improve VTO performance. At the same time the Marines removed the F.95 camera, RAT, and ARC-150 radio.

At the same time the Navy Air Rework Facility, where the modifications were accomplished, figured out how to increase the fatigue life of the airframe (something Hawker Siddeley had been unable to do economically), and extended the Harrier's life from 3,000 hours to 4,000. The total modification package only added 197 pounds. The first modified aircraft (BuNo 158384) made its maiden flight on 5 May 1979.

The modified aircraft were redesignated AV-8C, and began reaching operational units in 1983. All Marines squadrons received a small number of AV-8Cs, and VMA-513 was completely reequipped with the new model. But the AV-8C had a short service career, all being withdrawn from Marine Corps service (along with the remaining AV-8As) by the end of 1987.

Spanish Matadors

With a bit of political intrigue, Spain became very interested in the Har-

Unusually, this AV-8C appears to be in a solid color paint scheme instead of the original two-tone upper surfaces. Like most AV-8A/Cs in the early 1980s, the colorful national insignia has given way to a much smaller subdued insignia, still on the side of the air intake. The antennas for the ALR-45F radar warning receiver can be seen on the wingtip (a small light-colored protrusion from the leading edge) and on the tail stinger (a small protrusion on the extreme end). The APR-43 antenna would normally be mounted immediately ahead of the nose landing gear door, but does not appear to be fitted to this aircraft. (Michael Grove via the Mick Roth Collection)

During hover the nozzles are pointed straight down, allowing the Harrier to balance on a column of thrust. Reaction control valves located in the nose, tail stinger, and each wingtip provide attitude control during nonaerodynamic flight. The reaction controls use bleed air from the fan section of the Pegasus engine. (Mick Roth)

AV-8A (BuNo 158960) from VMA-513 photographed at MCAS Beaufort, SC, on 8 July 1974. The two small UHF antennas behind the cockpit were shared with the RAF's GR1/3, even though the radio attached to them was different. Another UHF antenna is flush mounted on the lower forward part of the tail. (Jim Tunney via the Mick Roth Collection)

Here is an AV-8A in subdued markings. All eight of the pressure relief doors are open, as is the canopy (probably because the photo was taken at NAF Fallon where it is hot, and the Harrier's air conditioning was marginal, at best). A pair of Aden 30-mm cannon pods and an ALE-37 chaff dispenser pod are on the fuselage, while the inboard wing station carries a napalm container. (Mick Roth)

Access panels have been removed from this Aden 30-mm cannon pod, showing some of its internal components. An ALE-37 chaff dispenser pod is mounted on the centerline station. Noteworthy are the reference markings around the "cold" nozzle – 10° and 55° are highlighted since these are the normal cruise and STO settings, respectively. The nozzle could rotate slightly forward to allow flight to the rear, although this was seldom done. (Mick Roth)

rier during 1972. The Spanish and British governments were not on particularly good terms at the time, mainly over the status of Gibraltar, so Spain could not order the aircraft directly from Hawker Siddeley. Things became almost comical in hindsight. A Spanish order for navy frigates from a British shipyard had been summarily cancelled by a new British government, further straining the relationship. Nevertheless, Hawker Siddeley had arranged for a demonstration of the Harrier aboard the Spanish aircraft carrier SNS *Dédalo* (the former USS *Cabot*). Then the Spanish government refused to allow the Harrier to operate from, or to overfly, Spanish territory. Amazingly, the Spanish Navy still wanted the demonstration, and arranged for the *Dédalo* to move from her normal Mediterranean station to a location off the coast of Portugal. Hawker pilot John Farley flew a GR3 from the UK non-stop to the *Dédalo*. Despite some fears that the Harrier would ignite the ancient wooden deck on the Spanish carrier, the demonstration went off perfectly. But Spain would still not order the aircraft from Britain, and it is unlikely Britain would have accepted the order in any case.

Although the VA.1s began life identical to the AV-8A, their upgrades more closely parallel the GR1, although they never received the extended LRMTS nose. The British-style RWR antennas are located on the leading edge of the vertical stabilizer and the end of the tail stinger. Oddly, the vertical has the extended fin cap developed to compensate for the LRMTS nose, although the nose on this Harrier is still original. Interestingly, the wing on this Harrier appears to be from a different aircraft—note the aircraft number on the air intake and compare it with the one on the upper wing surface. (Nick Challoner)

The Spanish Harriers wear a light grey upper surface with an off-white bottom. National markings are full-color red and yellow roundels. Like the AV-8A/C, the VA.1s have received formation light strips on the forward fuselage and vertical stabilizer. The RWR antennas are connected to a Marconi Sky Guardian 2000 radar warning system. (Fournier Michel via the Mick Roth Collection)

Discussions with the United States solved the dilemma. Six additional AV-8As (BuNo 159557/562) were added to the final USMC order, and these aircraft were then resold to the *Armada Espanola* under the VA.1 Matador (Mk 54 by Hawker, AV-8S by the USMC) designation. Two TAV-8A two-seaters (BuNo 159563/564) were also procured under the VAE.1 designation (BAe Mk 59 or TAV-8S). These aircraft were essentially identical to the American AV-8s, with the inclusion of different radios that were compatible with the units used by *Armada Espanola* helicopters. The Matadors were equipped with essentially the same RWR as the GR3, and carry the same antenna package on the leading edge of the vertical stabilizer.

Five additional Spanish VA.1s (BuNo 161174/178) were ordered later, primary as attrition aircraft. The tensions between the Spanish and British governments had eased, and these were purchased directly instead of being diverted through

This is how the Matadors looked immediately after delivery. The vertical stabilizer is devoid of RWR antennas, and a USMC-style VHF/FM antenna is located on top of the fuselage. A broad-blade UHF antenna is also located on top of the fuselage, something the American Harriers did not have. This was part of the unique radio equipment installed in the VA.1 for compatibility with Armada Espanola *helicopters.* (Mick Roth Collection)

the United States. Although these aircraft were essentially identical to the earlier aircraft, they were called Mk 55s instead of Mk 54s. The engine was identical to the 21,500 lbf Mk 103, but was called the Mk 150 by Rolls-Royce.

Spanish pilots underwent conversion training in the US during 1976, with the unfortunate loss of one of the AV-8S aircraft. The remaining aircraft were later ferried to Rota, Spain, where they were operated by No. 008 Squadron of the Spanish Naval Air Arm from the *Dédalo*. This made the Spanish Navy the first in the world to operate a V/STOL aircraft. When it came time to replace the aging *Dédalo* in 1977, an order was placed with the Bazán shipyard for a carrier based on a stillborne US Navy sea control ship design. The resulting *Principe de Asturias* is a 16,700-ton aircraft carrier equipped with a 12° ski-ramp on its bow. It was launched in 1982, performed sea trials in 1987, and was commissioned in 1988.

The Spanish Harriers were wired to carry Sidewinders on the outer wing pylon, although a training round is being carried here. (British Aerospace)

When Spain began to acquire more modern AV-8Bs, the Matadors were placed in a supporting role at Rota, mainly for training and indoctrination. In early 1992, a $70 million deal was signed by the Spanish to transfer seven single-seat and a pair of two-seat aircraft to the Royal Thai Navy. At the same time, a 11,500-ton carrier named *Chakkri Naruebet* was ordered from the Bazán shipyard. Although intended primarily as a helicopter carrier operating Sikorsky S-70B Seahawks, it is equipped with a 12° ski-ramp at the bow. The Harriers were refurbished prior to being transferred to Thailand, although they still wear their light-grey and white paint schemes.

The Royal Thai Navy has acquired several of the ex-Spanish AV-8As. This was the second aircraft delivered to Spain (BuNo 159558). The aircraft have not changed much from their final Spanish appearance, although it is interesting to note that they carry "AV-8A" and their USN BuNo on the ventral stabilizer. (Andy Moulton via Kev Darling/BBA)

The early Harriers used three distinctly different cockpits. The AV-8A (above left) was the simplest, and was generally similar to the GR1 (above right) except the Ferranti moving map display (large display on center panel) has been replaced with a standard gyro compass. Neither of these aircraft has been fitted with radar warning receivers. (British Aerospace)

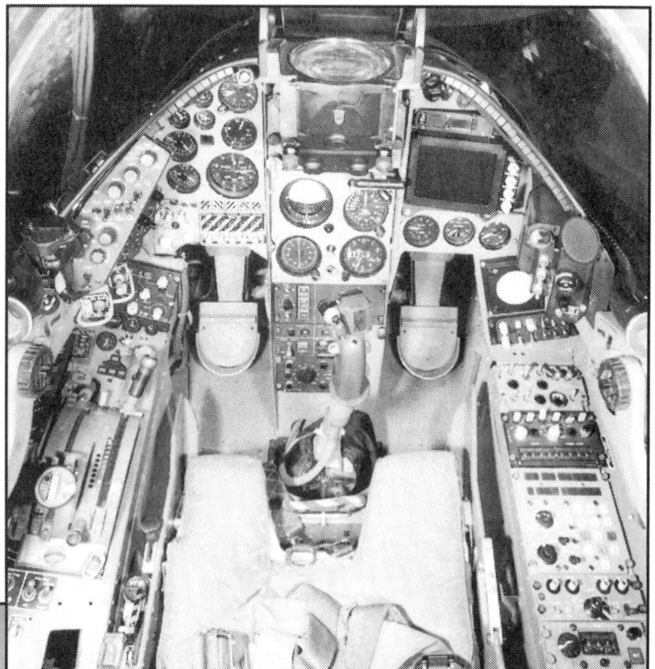

The Sea Harrier (right) had the most sophisticated cockpit, with the large display for the Blue Fox radar occupying the upper right panel. The white circular scope below and to the right of the radar is the RWR display. The throttle is on the extreme left console, with the nozzle control immediately to its right. (British Aerospace)

There were several unique features to the Harrier. A large yaw vane was mounted in front of the windscreen. This measured yaw during hover and extreme low-speed flight. After an unfortunately accident where a pilot was killed ejecting through the canopy, miniature ("mild" in US documentation) detonating cord was glued to the inside of the canopy. This cord is detonated a few milliseconds prior to the seat ejecting, ensuring the canopy shatters more completely. (Mick Roth)

6 SEA HARRIERS

HARRIERS FOR THE ROYAL NAVY

Almost a year prior to the original P.1127's first flight, engineers at Hawker had prepared a report detailing the possibilities of using a vectored-thrust V/STOL aircraft in a naval role. In this case the engineers provided data that indicated the new aircraft could operate, not just from aircraft carriers, but from frigates, merchant ships, and possibly even assault landing craft. Although no action was taken as a result of this report, the idea slowly began to take hold. The Admiralty began drafting a requirement in 1962 for an advanced V/STOL carrier-borne interceptor to replace the de Havilland Sea Vixen. This requirement led directly to the first tests of a V/STOL aircraft aboard a ship when XP831 landed aboard HMS *Ark Royal* on 8 February 1963.

These tests were sufficiently successful for the Royal Navy to support a maritime version of the proposed P.1154 supersonic V/STOL aircraft. But the P.1154 project ran into trouble and the RN opted to purchase Spey-powered versions of the F-4 instead. To go with their new aircraft, the Royal Navy planned to build a large-deck conventionally-powered aircraft carrier similar to the Forrestal-class carriers being built by the US Navy. The British government had other ideas, and the large-deck carrier program was cancelled in 1966.

The decision at this time was that the Fleet Air Arm would continue to operate their Phantoms and Buccaneers from existing ships until they were retired. The last large RN carrier, the HMS *Ark Royal*, would be withdrawn from service in 1978. After that, the RN would be limited to helicopters at sea, and air cover for surface ships would be provided by fixed-wing aircraft operating from land bases, where they were available.

Therefore, during the late-1960s the RN developed a new class of

The Sea Harrier FRS1 was the first Harrier to use a raised cockpit, and also the first to feature radar. Unlike the RAF's Harriers, the Sea Harrier carried radar warning equipment from the beginning and has always had the RWR antenna fairing on the leading edge of the vertical stabilizer. Also unlike the RAF, the RN specified that the Sea Harrier would be capable of carrying the AIM-9 Sidewinder.(Mick Roth Collection)

Besides looking sharp, the original Sea Harrier gloss blue-grey and white paint scheme assisted in keeping the aircraft clean, thereby eliminating some of the corrosive effects of the sea spray. (Royal Navy via the Terry Panopalis Collection)

"through-deck cruisers" intended to operate antisubmarine helicopters, and the first ship of the type, HMS *Invincible*, was laid down at Barrow-in-Furness during 1972. Around the same time the RN requested permission to develop a version of the Harrier to provide fleet air defense from this new class of ship (which would soon be renamed "Command Cruisers"). Two RAF GR1s were tested aboard the 50,000-ton HMS *Eagle* during March 1970. They made 58 sorties, including night operations, with special equipment to provide at-sea alignment of their inertial systems. Based on the results of these trials the maritime version of the Harrier was approved in May 1975. The new FRS1 aircraft was subsequently named Sea Harrier.

The initial order was for three development aircraft, to be followed by 31 production versions. A single T4A two-seater was also funded, although the aircraft was delivered to the RAF. The small number of aircraft was justified by the fact that only six aircraft would be embarked aboard each of the three Command Cruisers. The remaining aircraft were for training, attrition, and for use while others were undergoing heavy maintenance, etc. Additional orders brought the eventual total to 57 FRS1s and three T4Ns, the navalized version of the T4A trainer.

The Sea Harrier was designed to fulfill fighter, reconnaissance, and strike roles (hence its FRS designation). Since the Invincible-class ships were not equipped with catapults or arresting wires, the standard mode of operation was short or vertical takeoff, and verti-

The three paint schemes used by Harriers during the Falklands are illustrated here. The dark and medium grey schemes (top and middle) were attempts by the Royal Navy to find an effective paint scheme for the South Atlantic, while the RAF's GR3 continued to use their traditional camouflage. (Terry Panopalis Collection)

cal landing. Equipment was limited to what could be incorporated into the existing Harrier design with minimal modifications. As it turned out, roughly 90% of the airframe, powerplant, and mechanical systems were common with the RAF Harrier, but over 90% of the avionics were unique to the Sea Harrier.

Another innovation would accompany the Sea Harrier. Independently, both RN LtCdr. D. R. Taylor and Hawker's Ralph Hooper conceived the "ski-ramp" launch system. This ingenious idea incorporated an upward curve in the forward end of the flight deck so that the Harrier would be launched in an upward ballistic trajectory when executing a STOL takeoff. The system allows the Sea Harrier to takeoff with a larger payload than either a conventional flat takeoff (without a catapult, which the Sea Harrier can not use, and the British cruisers do not have) or a vertical takeoff. A series of ski-jump trials was conducted during 1977 from a variable-angle ramp at the RAE Bedford using both a GR3 and the Hawker G-VTOL demonstrator. Various ramp angles were tried, with angles between 7° and 12° determined to offer the best performance while keeping risk within reason.

The first Sea Harrier (XZ450, seen here) made its first flight on 20 August 1979. Less than two weeks later it was making daily ski-ramp takeoffs at the Farnborough Air Show. (British Aerospace)

The first of the new Command Cruisers, the HMS *Invincible* (R05, deck letter 'N') was launched on 3 May 1977, having been somewhat delayed by the last minute decision to include a 7° ski-jump ramp on the bow. The ship began its sea trials on 26 March 1979, and was formally commissioned at Portsmouth on 11 July 1980. The HMS *Illustrious* (R06/L) was commissioned on 20 June 1982, and HMS *Ark Royal* (R07/R) was commissioned on 1 November 1985, equipped with a 12° ski-ramp. The *Hermes* was also modified with a ski-jump to support Harrier operations.

The basis for the Sea Harrier design was the GR3 in production for the RAF. The major change was a completely redesigned forward fuselage to accommodate a Ferranti Blue Fox radar. This radar was derived from the Sea Spray unit used in RN Lynx helicopters, but was significantly modified to accommodate both air-to-air and air-to-surface modes. The radar information was displayed on a TV-style display that also showed flight data in some modes. A larger Smiths Industries HUD driven by a more powerful computer was fitted. A new cockpit environmental system provided an output similar to that normally fitted to two-seat aircraft because of the additional demands of the radar equipment. As a result, the cockpit remained much more comfortable in tropical climates.

A self-aligning, two-gyro attitude reference platform and Decca 72 Doppler radar replaced the inertial unit used on the GR3 to avoid alignment problems at sea. The unit typically gave a navigation accuracy of 1.5 nm after a 50-minute flight. New radio navigation aids included UHF homing, TACAN, and an I-band transponder. Radio

The weather can cause some violent ship motions. Like most aircraft aboard ship, the Sea Harrier is usually well chained to keep it from moving unexpectedly. (Terry Panopalis Collection)

equipment included a multichannel PTR377 UHF/VHF unit and a standby D403M VHF-only unit. Unlike the RAF Harriers, an autopilot was incorporated that provided simple height, heading, and attitude hold. Handling qualities in V/STOL mode were improved by providing increased roll reaction control power, and allowing 2° more nose-down pitch control via greater horizontal stabilizer positive travel (leading edge up).

All electrical wiring was redesigned to be more corrosion resistant, along with self-sealing plugs. Special attention was given to running wiring bundles to avoid the electromagnetic interference typically found on ships. High-capacity hydraulic pumps, along with the additional hydraulic filters fitted to the last USMC AV-8A, were provided. Since magnesium is prone to corrosion in salt air environments, special attention was paid to eliminating as many magnesium components as possible. Almost all magnesium components in the airframe and engine were redesigned in various aluminum alloys. Only four magnesium components remained on Sea Harrier: the Pegasus gearbox (oil-washed on the inside, with non-air scrubbed outer surfaces); the nose wheel; and the two outrigger wheels. The wheels were retained because no excessive corrosion had been noted on RAF Harriers. To further aid in corrosion protection, the inside of all compartments was painted a gloss acrylic, and all ferrous engine components were coated with a sacrificial aluminum paint to retard corrosion. The revised (non-magnesium) Pegasus

The Sea Harrier has proven remarkably successful at launching and recovering aboard ship. The changes made to the Sea Harrier, namely replacing magnesium parts and special attention to corrosion resistant components, have resulted in an aircraft that needs little more maintenance than its land-based counterparts. (British Aerospace)

Mk 104 engine was essentially a navalized version of the Mk 103 with the same 21,500 lbf thrust.

The total weight penalty for the Sea Harrier conversion was less than 100 pounds, an achievement unequaled by any other modern land-based aircraft adapted for maritime use. One of the primary contributors of this lack of weight growth was the acceptance of vertical landing as the only means of shipboard recovery. Traditionally, arrested landings have required that the landing gear and associated structure be significantly strengthened to accommodate the tremendous forces associated with this type of landing. By accepting vertical landing as the only recovery method, it was decided that the existing Harrier landing gear could be used unchanged.

The Sea Harrier cockpit was completely redesigned to incorporate the multimode weapons system. Even during the early design studies it became clear that the RAF Harrier cockpit, which was largely unchanged from the original P.1127 that first flew in 1960, was ergonomically limiting. Since the forward fuselage was being redesigned in any case to accommodate the Blue Fox radar, the opportunity was taken to raise the pilot 10 inches, although overall width remained the same to avoid potentially disturbing the airflow into the intakes. Interestingly,

BAe engineers figured out how to use the existing windscreen assembly, as well as the existing canopy frame, although the transparency was "bubbled" to provide additional headroom and better visibility. New white (versus orange) night lighting was provided by edge-lit panels. A new Martin-Baker Mk 10 ejection seat replaced the MB Mk 9 used by the RAF, and provided a slightly faster ejection sequence (1.5 seconds versus 2.25).

Raising the cockpit allowed more equipment to be mounted under the cockpit floor, thus freeing yet more space in the extreme nose for the new radar. The first 5.3 feet of the nose folded to allow access to the radar equipment, and also to provide a smaller footprint for shipboard storage. The unit cost of a Sea Harrier was quoted at £6.8 million during the late 1970s, or about 18% more than the RAF's GR3.

The five weapon pylons were of an improved design compared to those used on the RAF Harriers, including stronger ejector release units. The pylons could accommodate a large range of RAF and USMC weapons. Like the USMC AV-8s, the Sea Harrier was given the capability to carry the AIM-9 Sidewinder on the outer pylons. A twin-rail launcher was later developed based on experience gained during the Falklands conflict. The inboard pylons could accommodate Martel and Harpoon antiship missiles. A F.95

The amazing Project SkyHook. The privately-owned G-VTOL two-seat demonstrator made several launches and recoveries using a land-based crane equipped with the special jib. Although no serious problems were encountered, no attempts were made to demonstrate this concept on a ship. (British Aerospace)

Arguably, the original FRS1 Sea Harrier was the sleekest-looking of the Harriers. This was largely ruined by the addition of the large radome for the Blue Vixen radar on this FA2 variant. However, most pilots will happily trade a slight degradation in appearance for the significantly improved capabilities offered by the new radar system. (Kev Darling/BBA)

strike camera was mounted in the right forward fuselage, but it proved nearly worthless in assessing bomb damage of Port Stanley during the Falklands campaign.

The prototype Sea Harrier (XZ450) made its first flight on 20 August 1978 at Dunsfold, followed quickly by three Development Batch (DB) aircraft. The first Sea Harrier unit, No. 700A Squadron of the Fleet Air Arm, was commissioned on 19 September 1979 with the delivery of XZ451. Between 24 October and 8 November five FRS1s and the G-VTOL demonstrator completed their type operational trials aboard HMS *Hermes* in the Irish Sea. No. 700A Squadron became the Sea Harrier Intensive Flying Trials Unit to continue operational testing of the Sea Harrier. The first carrier-deployed Sea Harrier unit was No. 800 Squadron aboard the HMS *Invincible*. The second Sea Harrier unit was No. 801 Squadron, and for a brief time during the Falklands conflict, No. 809 Squadron was commissioned operating Sea Harriers.

British Aerospace came up with an interesting concept to utilize Sea Harriers from almost any size ship without the need for any launching platform at all. A fairly large crane would be equipped with a special jib that could attach to the top of the Harrier's wing and fuselage. The aircraft would be lifted from its hanger (in many cases nothing more than a protective crate mounted on the deck) and swung over the side of the ship. The pilot would start the engine with the nozzles pointed downward while still attached to the jib and perform the normal preflight checkouts. The jib would release the aircraft which would enter a conventional hover then slowly transition to forward flight. Recovery involved the pilot matching the ship's speed beneath the extended jib and waiting for the crane operator to snatch the aircraft. Once the aircraft was securely attached to the jib, the pilot would shut down the engine and the aircraft would be lowered to the ship's deck.

This extraordinary concept was called "SkyHook" and the technique was demonstrated using a land-based crane and the G-VTOL two-seater. As far as is known, the concept was never attempted at sea, although the tests did not reveal any significant problems with either the hardware or techniques. In most probability it was decided that there was no need to operate Harriers from ships smaller than those that normally had helicopter pads.

A second, less exotic, technique for equipping merchant ships with Sea Harriers was the shipborne containerized air defense system (SCADS). In this concept a Sea Harrier and its support equipment would be carried in several modified cargo containers aboard large-deck container ships. The Sea Harrier container would be at the top of the stack of containers, allowing its top to open and the aircraft to fly out as needed to protect the ship or convoy. Although it sparked a great deal of publicity in the popular press, no direct applications for the idea could be found.

Falkland Islands

Operation Corporate focused the World's attention on a small island off the coast of South America, and also on the Harrier. It was the GR3 and the FRS1 that were the first Harriers to see combat during April-June 1982. Four GR3s from the RAF's No. 1 Squadron were ferried across the Atlantic with numerous in-flight refuelings and a brief stop on Ascension Island. When the RN task force sailed from Portsmouth on 5 April 1982, a total of 28 Sea Harriers were available for service, and 24 of these were embarked aboard HMS *Hermes* and Invincible. Six additional GR3s, along with the remaining four Sea Harriers, were placed aboard the merchant ship MV *Atlantic Conveyor*. During the trip across the Atlantic, at least one Harrier was kept on ready-alert aboard the Atlantic Conveyor. Once they arrived near the Falklands, all of the aircraft were readied, then flown to the carrier HMS *Hermes*, just nine days before an Argentinean Exocet sank the *Atlantic Conveyor.*

The first attack made by the Sea Harriers was a dawn low-level raid on the airfield at Port Stanley on 1 May 1982. Nine aircraft used cluster bombs and their cannon to disrupt operations at the airfield. At the same time other Sea Harriers attacked a small airfield at Goose Green, 50 miles to the west. The Argentineans were airborne with a mixed force of Mirages, Skyhawks, and Canberras, and the Sea Harriers managed to shoot down one of each type without losing any of their own.

During the assault on the Falkland Islands, the RAF Harriers were tasked with ground support duties, mainly attacking airstrips and other targets around Port Stanley and Darwin with cluster bombs. Four of the GR3s were shot down by ground fire, but all of the pilots escaped serious injury.

After the RAF Harriers arrived the Sea Harriers were largely used in the combat air patrol role since they were equipped to launch AIM-9 Sidewinders and the RAF aircraft were not, at least initially (the RAF hurriedly adopted such a capability). The FRS1 proved remarkably successful in this role. On 21 May the British made a major amphibious landing at San Carlos Bay and the Sea Harriers engaged Argentine Mirage IIIEAs, A-4P Skyhawks, and Aermacchi MB-338 ground attack aircraft. Reportedly, 23 Argentine aircraft were shot down against the loss (to ground fire) of one Sea Harrier and five British helicopters. A few days later, four Mirage IIIEAs were encountered by a flight of two Sea Harriers – all four Mirages were destroyed.

During the war, both RAF and RN Harriers received hasty installations of the Tracor ALE-40 chaff dis-

FA2 Sea Harrier from No. 899 Squadron at RNAS Yeovilton. The Sea Harrier continues to use the fixed refueling probe instead of the retractable unit used by the Harrier II family. The probe is normally removed from the aircraft since it contributes a fair amount of base drag, but it is installed whenever a long ferry flight is scheduled. The RN has no indigenous aerial refueling capability (i.e., no tankers). (Nick Challoner)

One of the FA2 trials aircraft during July 1992. A Sidewinder acquisition round is on the outboard pylon. Noteworthy is the the pitot tube beneath the RWR antenna on the vertical stabilizer. The pitot was moved from its nose location to the tail during the FA2 conversion, although normally the pitot is much longer. (Nick Challoner)

penser. At the same time, Marconi Sky Shadow jamming pods (the same as normally carried by Tornado GR1s) were repackaged into the shells of Harrier cannon pods, allowing the Harrier to carry a jammer without taking up a valuable weapons pylons (the cannon not being used too often). There are mixed reports as to whether these systems were actually used by Harriers during the Falklands. Modifications were also made to the RAF Harriers to launch the AIM-9, using the same pylons originally designed for the USMC.

By the end of the Falkland campaign, the 28 Sea Harriers had flown more than 2,000 sorties with an overall reliability rate of over 90 percent. Six Sea Harriers were lost, two to ground fire, and four to accidents, including two that collided while flying through clouds during the trip to the Falklands. No Sea Harriers (or RAF Harriers) were lost in air-to-air combat.

The success of the Sea Harrier during the Falklands led to an order for 14 additional aircraft. This provided attrition aircraft for those lost during the war, and also allowed the RN to boost the normal Sea Harrier complement from eight aircraft to ten per squadron.

F.A. Mark 2

Although the Royal Navy was generally pleased with the performance of the Sea Harrier during the Falklands campaign, the conflict did point out some shortcomings that were not totally unexpected. The major problem was the relatively limited performance of the Blue Fox radar. What was needed was a true look-down/shoot-down radar that incorporated a track-while-scan capability. In 1983 it was decided to replace the Blue Fox radar with the advanced multimode pulse-doppler Blue Vixen radar, along with the capability to fire the new AIM-120 AMRAAM medium-range missile. Up to four AIM-120s can be carried, two replacing the normal Sidewinder installation, and two on pylons that replace the cannon pods. At the same time the ability to launch the BAe Sea Eagle sea-skimming antiship missile and ALARM (Air-Launched Anti-Radar Missile) was added.

The modified aircraft were redesignated F.R.S. Mark 2, and the first two aircraft were modified beginning in 1985. This designation was later changed to F.A. Mark 2 (Fighter Attack). The most notable external changes are a revised radome shape, and a 13.75-inch plug in the fuselage behind the wing to provide an extra equipment bay. This was accomplished by installing an entirely new rear fuselage instead of just extending the existing fuselage, mainly because numerous hairline cracks were found in the original structures. Subtle changes were also made to the inner one-third of the wing, including an additional fence, a kinked leading edge, and the deletion of the dog-tooth and one of the 12 vortex generators.

Two BAe 125-600B business jets were modified for use as an avionics testbed for the Blue Vixen radar. A Mil-Std-1553B data bus was included to link all the advanced avionics and weapons systems and a Marconi Sky Guardian ECM system replaced the original ARI.18233 system.

The ram air turbine fitted on the upper fuselage was deleted since it offered little real advantage in an aircraft that had a one-in-three glide angle and extremely hazardous dead-stick landing characteristics. Pilot's were advised to use the Martin-Baker Mk 10H ejection seat if an engine restart does not work.

The Blue Vixen radar was first tested on a BAC-111, but was soon transferred to a BAe 125 (XW930). A second BAE 125 (ZF130, shown above) was fitted with both the Blue Vixen radar, a complete weapons system, and an underwing pylon to carry AIM-9 and AIM-120 missiles. A partial FA2 cockpit, including the HUD, stick, and some instrumentation, has been installed in the normal co-pilot's location. (Nick Challoner)

The FRS1 had proven itself during the Falklands conflict to be a formidable air-to-air fighter using the AIM-9 Sidewinder. The FA2 has the capability to use the new AIM-120 AMRAAM, greatly improving its aerial capabilities. A new set of pylons can replace the fuselage strakes (or gun pods) to carry a second pair of AIM-120s. The AMRAAM is marginally too heavy to allow carrying two on on the outer pylons. A mixed load of AIM-120s and Sidewinders can also be carried in any combination. (British Aerospace)

The engine for the FA2 was a 21,500 lbf Pegasus Mk 106 variant of the F402-RR-406 installed in the AV-8B. Like the original Sea Harrier engine, the major changes to this powerplant involve substituting conventional alloys for the extensive magnesium normally used.

Considerable operational benefit was gained by the ergonomic integration of additional switches as part of the control column and throttle handle functions. These hands-on throttle and stick (HOTAS) controls provide simultaneous control of the aircraft, radar, and weapons system without the need to operate separate controls and switches. This allows the pilot to fully concentrate on the total tactical situation presented on the multifunction display and HUD.

The first FA2 (ZA195) flew on 19 September 1988, with the second (XZ497) flying on 8 march 1989. On 7 December 1988 a contract was awarded for the conversion of 31 FRS1s between 1991 and 1994. On 6 March 1990 an order for ten new-build FA2s was announced, mainly to cover attrition losses. The Blue Vixen radar made its first Sea Harrier flight in the second FA2 prototype (XZ439) on 24 May 1990. Carrier compatibility tests were conducted between 7 and 16 November 1990, and AMRAAM firing trials began in late 1991.

The relatively large diameter of the Blue Vixen radar nose on the FA2 Sea Harrier is readily apparent. The radome on the big-wing AV-8B+ is similarly sized, although the Blue Vixen is arguably the more capable than the USMC's APG-65. (Nick Challoner)

India was the only export country for the Sea Harrier. Interestingly, it operates them off an ex-British aircraft carrier. A total of 23 Sea Harrier Mk 51 and four Harrier T60 (plus two ex-RAF T4s) two-seaters were purchased. (British Aerospace)

Indian Sea Harriers

India is the only other nation to purchase the Sea Harrier, although Italy considered the aircraft before deciding on the AV-8B instead. The Hawker two-seat demonstrator (G-VTOL) first flew from the Indian carrier INS *Vikrant* in July 1972, but it took until December 1979 for six Sea Harrier Mk 51s and two Harrier Mk 60 trainers to be ordered. The Mk 51s were essentially the same as the FRS1, while the Mk 60 two-seater was based on the T4N. Major changes in the single-seaters included a slightly detuned radar, a gaseous oxygen system instead of liquid oxygen, different voice and IFF radios, and the substitution of Matra R.550 Magic air-to-air missiles for the normal US-made AIM-9 Sidewinders.

The 34th Sea Harrier manufactured was the first Indian aircraft, and made its first flight on 6 August 1982. Ten additional Mk 51s and a single Mk 60 were ordered in November 1985. The former HMS *Hermes* was transferred to the Indian Navy as the INS *Viraat*, and in order to equip the new carrier, an additional seven Mk 51s and one Mk 60 were ordered in August 1986. Two ex-RAF T4s were purchased in late 1996 to replace two Mk 60s that were lost in accidents Surprisingly, the US offered to sell India five TAV-8As that were in storage. If India had elected to purchase these (they did not, mainly due to the costs of refurbishing the aircraft to Mk 60 standards), they would have been the first combat aircraft ever sold to India by the United States.

The 22 Indian Sea Harriers and a pair of two-seat trainers remaining in service as of mid-1998 are considered sufficient since the *Vikrant* was decommissioned in 1996. The Indian Navy expressed an interest in obtaining the FA2 upgrades, but the US government declined to issue an export license for the AIM-120 AMRAAM missiles. An alternate modification program began when Israeli Aircraft Industries (IAI) offered a relatively affordable set of upgrades that includes the Elta ALM/2032 pulse-Doppler radar integrated with the BAe ASRAAM missile and Elta 8420 ECM suite. An updated cockpit and various structural modifications are also included in the IAI proposal, which is designed to keep the Indian Sea Harriers operational until the year 2010.

7 Two-Seat Harriers

Small-Wing Solutions to Unique Training Requirements

At the same time the original P.1127(RAF) design effort began, a feasibility study of a two-seat derivative was undertaken. Four alternate designs were investigated, including conventional tandem and side-by-side seating, a twin front fuselage where the student and instructor were in separate cockpits on either side of a central engine intake, and a design where the instructor's cockpit was immediately forward of the vertical stabilizer. The driver was to provide a conversion trainer to assist pilots in coming to grip with the realities of vertical flight. But this had to be accomplished without upsetting the delicate balance so critical to the Harrier's V/STOL performance, and without costing a great deal of time or money. Designing a two-seat derivative of a conventional aircraft only requires maintaining a satisfactory relationship between the aircraft's center of gravity and its aerodynamic center. A V/STOL aircraft also dictates that the thrust center while in vertical flight be considered. In addition, the air intakes on the Harrier had been developed to a high state of efficiency, and any changes to the forward fuselage risked upsetting this. In the case of the Harrier, the designers managed to maintain the delicate balance, but produced a much less esthetic aircraft than the original single-seat Harrier.

On 4 January 1965 a proposal was submitted by Hawker, resulting in Air Staff requirement 386 being issued for a two-seat Harrier. It was not until late 1966, however, that a contract for two prototypes (XW174/175) was signed. The original cockpit and nose assembly were moved 47 inches forward, and a constant cross-section extension was added with a second cockpit raised 18 inches to provide the instructor with good visibility. Since the rear cockpit still intruded 7 inches aft of the previous cockpit bulkhead, the environmental equipment had to be relocated to a new fairing behind the cockpit. The air conditioning was also increased in capacity to deal with two pilots. The inertial platform and F.95 oblique camera were moved from the nose to a position below the rear pilot.

The large sideways-opening canopies of the TAV-8A provided easy access to the cockpits. The USMC waited until after it had received all of its single-seat aircraft before it ordered its two-seat trainers, an understandable but unfortunate sequence of events. (Bill Curry via the Mick Roth Collection)

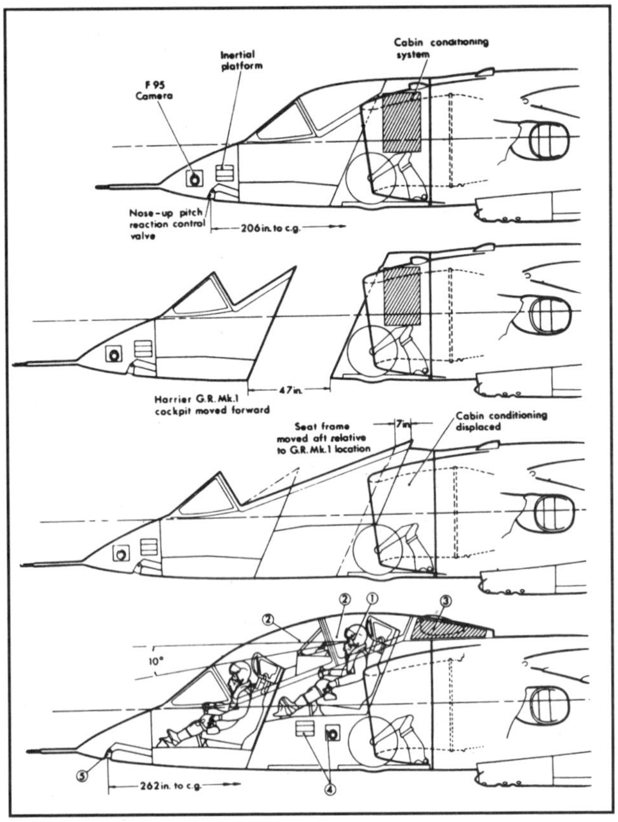

The evolution of the T2 forward fuselage from the GR1. (1) rear cockpit providing 10° downward view dead ahead; (2) new sideways-opening canopies; (3) cabin conditioning system enlarged and repackaged aft of canopies; (4) inertial platform and F.95 camera relocated under rear cockpit; (5) front reaction control valve moved forward. (British Aerospace)

To maintain the center of gravity, 170 pounds of ballast were installed in an enlarged tail stinger, resulting in an aircraft that weighed 1,450 pounds more than the single-seater. For wartime operations the rear seat and the tail ballast could be removed and the aircraft flown as a single-seater, although it still weighed 850 pounds more than the GR1. To provide satisfactory directional stability the vertical stabilizer was moved 33 inches rearward and raised 11 inches.

The first T. Mark 2 two-seater (XW174) made its maiden flight on 22 April 1969. Unfortunately, the aircraft was lost on 4 June 1969 when the fuel system malfunctioned. When test pilot Duncan Simpson ejected, a large piece of canopy struck his helmet, causing a neck injury. The Martin-Baker seats were designed to eject through the canopy (the thinking being that at low altitudes there was not sufficient time to jettison the canopy separately) using canopy-breakers integral with the top of the seats. This accident led to the development of miniature detonating cord (MDC—called "mild" detonating cord in US service) which was bonded to the inside of the canopy of all Harriers. This cord is detonated concurrently with seat ejection, and better fragments the canopy.

Flight tests with the second two-seater (XW175) revealed that directional stability was marginal at higher angles of attack. Various changes were investigated, including adding vortex generators to the rear canopy frame, repositioning the air conditioner intakes, and changing the upper contours of the air intakes, all to no avail. It was then decided to add an extension to the top of the vertical stabilizer. Extensions of 2, 18, and 23 inches were tested, and the consensus was that the 23-inch extension provided the best handling. Engineers worried, however, that the extension would exceed the structural integrity of the vertical, and it was subsequently decided that the 18 inch extension was the best compromise.

But the saga of the two-seater's vertical stabilizer was not over. In operational service it was felt that the extra vertical surface provided too much stability at low angles of attack. When the leading edge RWR antenna was added (providing

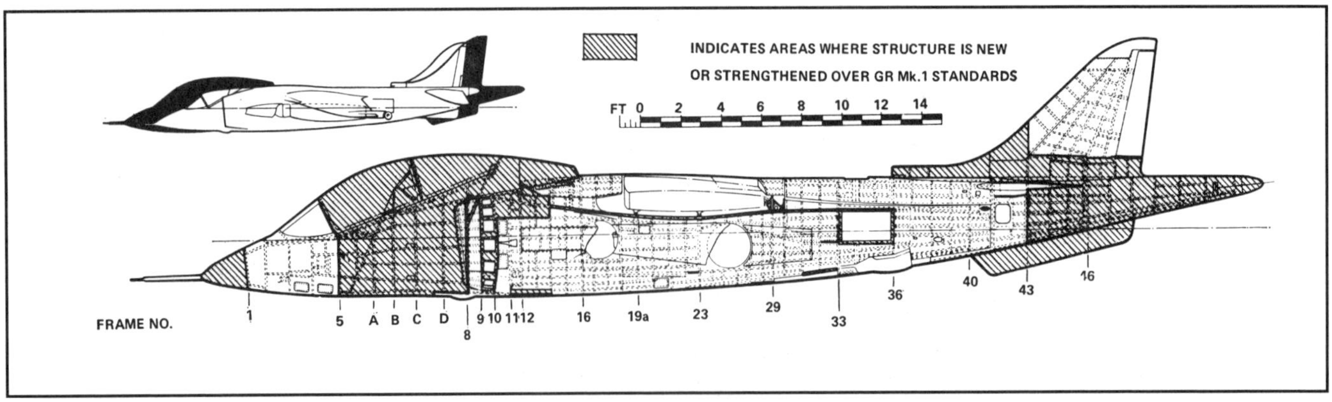

British law does not allow civilian aircraft to carry weapons, so the Hawker-owned G-VTOL demonstrator was assigned the RAF number ZA250 so that it could carry and fire live ordnance. The aircraft was frequently used to show the advantages of the ski ramp. (MoD via the Terry Panopalis Collection)

The T4 incorporates all the improvements made to the GR3, including the LRMTS nose and RWR antennas. Like all T4s, the original 18-inch fin extension was removed after the RWR antenna was added. (Kev Darling/BBA)

T4N from No. 899 Squadron at RNAS Yeovilton Since the Royal Navy did not use the LRMTS, its aircraft were not modified with the extended nose. (Nick Challoner)

This Spanish two-seater has worn a small shark mouth for most of its career. Although the Spanish single-seaters received RWR updates, the two-seaters did not. The vertical on this aircraft shows the original 18-inch extension. (Nick Challoner)

The beginning of a career, and an end. In August 1976 (above) the TAV-8A was just entering service. Twelve years later, another TAV-8A (right) is in storage in MASDC. (Ray Leader [above] and Craig Kaston [right] via the Mick Roth Collection)

more surface area), this became unacceptable.The 18-inch extension was removed and the aircraft would simply restricted from flying at high angles of attack.

The first two aircraft had used the Pegasus Mk 101 engine, and had been designated T2 by the RAF. The inclusion of the Pegasus Mk 102 into the next production aircraft resulted in the T2A. The inclusion of the Mk 103 engine, ARI.18223 RWR receivers, and LRMTS laser nose resulted in the T4. A few aircraft received the new engine and RWR receivers, but not the laser nose, and these were designated T4A, including one funded by the Royal Navy prior to their purchase of dedicated T4Ns. A total of 27 two-seaters (two T2 prototypes, ten production T2s, four T2As, seven T4s, a single T4A, and three T4Ns) were built. In addition, a single Mk 52 demonstrator (civil registry G-VTOL, but also carried the RAF number ZA250) was built as a private venture by Hawker Siddeley, and it made its first flight on 16 September 1971.

A new T. Mark 8 trainer was proposed in 1990, using the systems from the Sea Harrier FA2 along with the normal T4 airframe. The three existing T4Ns were converted to the new standard, and four new-build T8s were manufactured and assigned to No. 899 Squadron at Yeovilton.

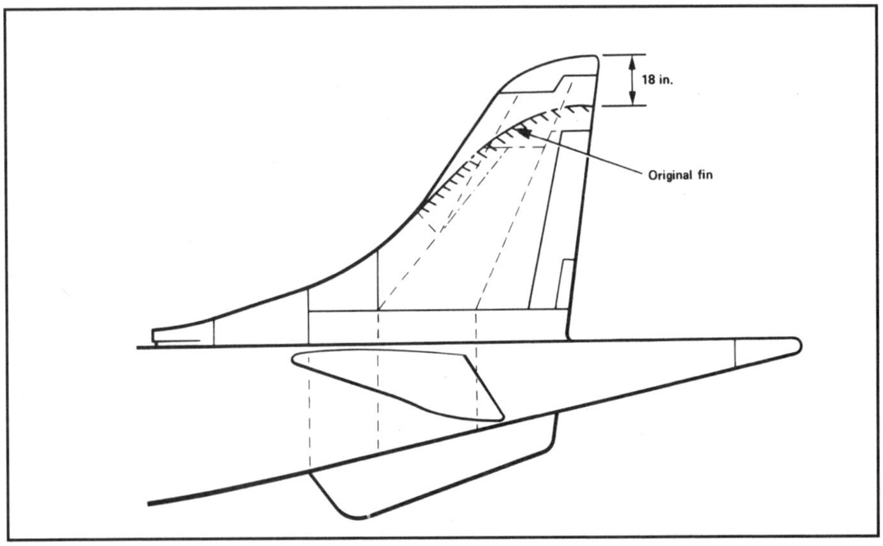

Multinational Service

Kestrels and Harriers in Color

The Harrier has operated in a great many paint schemes over the years. Interestingly, although the aircraft has operated in the armed forces of six different countries, most of the paint schemes originated with either the British or the Americans.

Early Harriers (GR1 and AV-8A) were both operated in standard RAF tactical camouflage of grey and green.

Perhaps the most painted Harrier was the Hawker-owned two-seat demonstrator (G-VTOL) which received new paint frequently in its role as a sales demonstrator.

When the Royal Navy received their Sea Harriers, they were originally in the then-standard RN scheme of dark grey upper surfaces and white lower surfaces. The Indian Navy painted their Sea Harriers similarly. The Falklands gave birth to an overall light grey scheme for the Sea Harriers, although this was later replaced by an overall dark grey.

All the big-wing Harriers have been delivered in various shades of grey camouflage, and there seem to have been many variations to the theme. The latest is the standard American two-tone grey. RAF GR5/7s operate in overall dark green.

Various test aircraft and "specials" have exhibited very colorful schemes from time-to-time. Unfortunately, limited space prevents showing more than a small sample.

One of the Hawker NASA XV-6 Kestrels assigned to the Langley Research Center in Hampton, Virginia. This photo was taken on July 1974. (Dave Ostrowski via the Mick Roth Collection)

After the Marine Corps retired their remaining early Harriers, NASA received this AV-8C to use as a test aircraft into V/STOL concepts. NASA also operates one of the AV-8B prototypes. (Mick Roth Collection)

When the Marine Corps purchased the Harrier as an "off-the-shelf" system, it included the standard RAF paint scheme. This early (1973) example shows the full-color national insignia that was used during the early part of the AV-8's career. (Ben Knowles via the Mick Roth Collection)

The first full-scale development AV-8B wore this unusual paint scheme during 1982 tests. Noteworthy is the size of the strakes (LIDs) under the fuselage. (Michael Grove via the Mick Roth Collection)

The original Spanish Harriers were delivered in a light gull grey and white paint scheme developed by the US Navy (which ordered them from Britain for Spain). The VA.1s wore the full-color roundel for their entire careers. (Kev Darling/BBA)

Standard RAF Harrier GR1 (XW766) in early 1975. Full-color roundels and fin flash would later give way to subdued markings. This aircraft was assigned to No. 3 Squadron. (Kurt Thomson via the Mick Roth Collection)

Dark grey scheme used by the Sea Harrier after the Falklands. While participating in the Falklands campaign, most FRS1s used a much lighter shade of grey. Subdued markings still use red and blue roundels. (Mick Roth Collection)

The original blue-grey and white scheme used on Royal Navy (and Indian Navy) Sea Harriers. Note full-color national insignia. (Rolls-Royce via the Terry Panopalis Collection)

An unusual paint scheme adorns this VMA-231 AV-8B (BuNo 162734) on 12 December 1987. (Craig Kaston via the Mick Roth Collection)

An AV-8B (BuNo 163181) from VMA-311 shows the standard paint scheme used by the Marine Corps in late 1991. (Craig Kaston via the Mick Roth Collection)

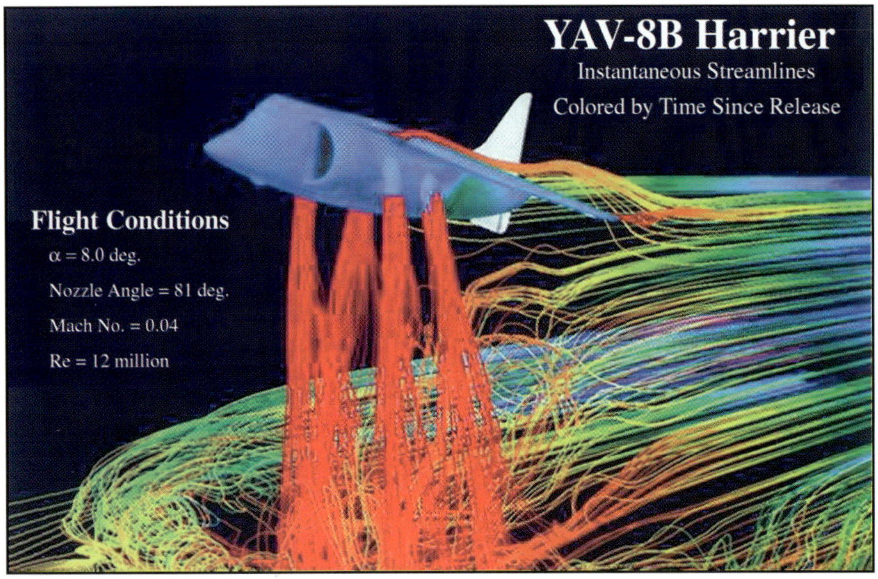

Computational fluid dynamics image of the instantaneous streamlines generated by a model of the YAV-8B. Noteworthy is the small cockpit on the model. The four exhaust streams are pointed straight down, as if the aircraft was in hover. The image was made on 11 May 1992 at the NASA Ames Research Center. (NASA)

8 Super Harrier

McDonnell Douglas Invents the Big-Wing Harrier

In the early 1970s, Bristol tested a demonstrator engine called Pegasus 15. This 24,500 lbf engine used a fan section that was 2.75 inches larger in diameter than the Pegasus 11. The larger fan diameter meant that the engine could not be fitted into existing Harriers without major modifications, something judged uneconomical by all involved. The original licensing agreement between Hawker and McDonnell Douglas covered development of "derivative" aircraft, and in 1972 a joint US/UK Advanced Harrier program was begun to investigate aircraft using the new engine. A 26-volume project definition document was issued on 13 December 1973.

The Advanced Harrier was intended to meet the requirements being generated by the US Navy for a high-performance attack aircraft system (HIPAAS), intended to replace the McDonnell Douglas A-4 Skyhawk as well as the first-generation AV-8A Harriers. The aircraft was also intended to fulfill an RAF requirement for a Harrier replacement, and to serve aboard the Royal Navy's Command Cruisers (remember, the Sea Harrier had not been developed yet). In the United States this program was known as the "AV-16," although the designation was decidedly unofficial. Hawker Siddeley apparently called the aircraft the P.1184 (although there is some controversy over this).

The American designation illustrated the intended purpose of the aircraft – to double the payload and range capability of the existing Harrier. The designers did not truly succeed, although they developed a useful improvement over the first generation aircraft. Specifically, the aircraft would have a VTO gross weight of 21,500 pounds and a STO (320-foot roll) weight of 25,000 pounds. On a 345-mile mission the aircraft could lift 2,000 pounds of weapons in VTO mode, or 4,000 pounds in the STO mode.

The AV-16/P.1184 featured the new Pegasus 15 engine, which dictated a larger diameter fuselage. Also included were a supercritical wing

Although they bear an obvious family resemblance to earlier Harriers, the AV-8B and other big-wing Harriers are essentially a new design. About all that remains of the original airframe design is the center fuselage. The new Harrier is a true multinational effort, with the Americans and British taking the lion's share, but Italy and Spain also contributing parts. (Mick Roth Collection)

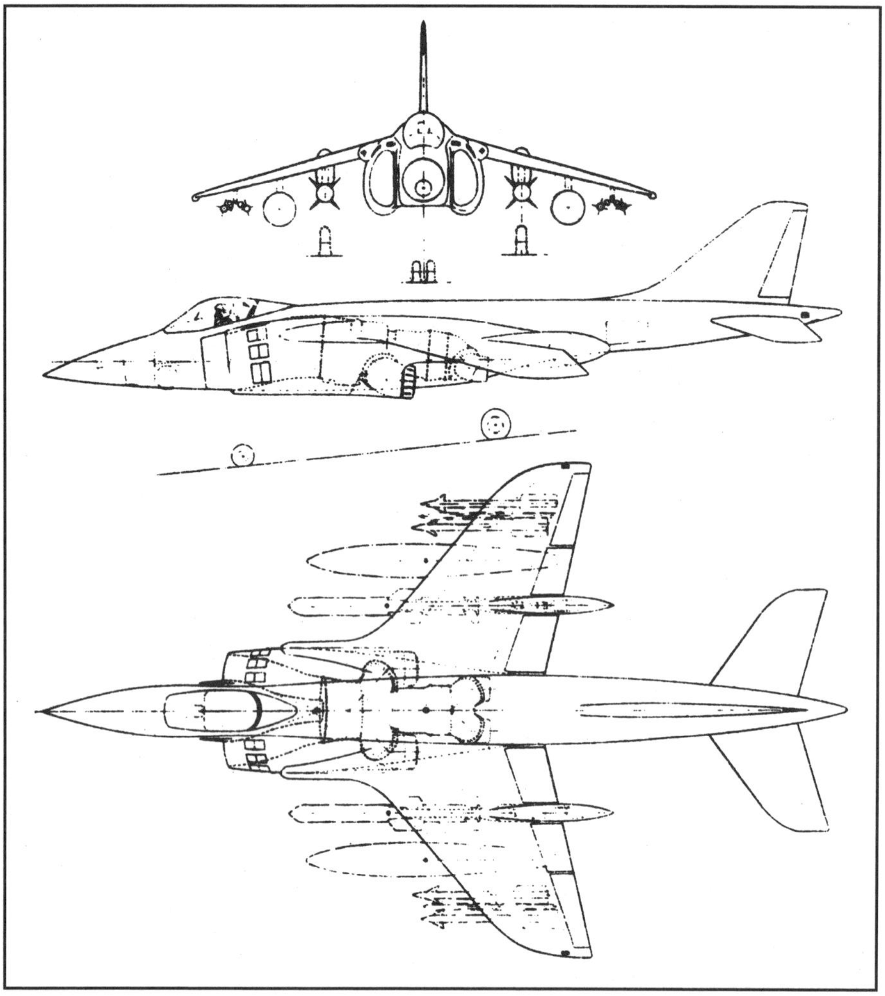

General Arrangement of the proposed supersonic AV-16-S6 using a PCB-equipped version of the Pegasus 15 engine. (McDonnell Douglas)

equipped with six weapons stations, a raised cockpit similar to that used on Sea Harrier, strengthen undercarriage, and new avionics. Nevertheless, the base AV-16 remained a subsonic aircraft, albeit one with a much improved payload capacity. However, a second version, the AV-16-S6 (P.1185), employed a PCB-equipped version of the Pegasus 15 and would be supersonic at high altitude.

The British government pulled out of the program in early 1975 due to ever decreasing defense budgets. Although the Pegasus 15 demonstrator had been tested, Bristol estimated that it would cost $250 million to finish development of an operational version of the engine. Total development costs for the AV-16 were estimated at over $2 billion, or even higher if the supersonic AV-16-S6 were developed also. The United States was unwilling to fund the effort alone, so in mid 1975 the program was cancelled. This did not overly disappoint the Marines, who were worried that the proposed aircraft was rapidly becoming too complex for the mission they flew.

McAir discontinued work on the Harrier derivatives, but Hawker took a different approach. Since the Pegasus 15 engine was too large to fit into the existing Harrier airframe, the engineers at Kingston abandoned it and tried to figure out ways to improve what they already had. Investigations were begun on a new wing that could be retrofitted to the existing aircraft to increase performance. Although some progress was made, calculations showed that the performance gains were not going to justify the development costs, and the project was cancelled.

But McAir saw some benefit to the research accomplished by Hawker. McAir engineers figured that a Harrier with a larger-area (230 versus 201 sq ft) supercritical wing manufactured from carbon-fiber composites and equipped with large slotted flaps, could carry 2,000 pounds more fuel. Larger air intakes equipped with two rows of pressure-recovery doors (16 total) could increase the maximum thrust of the Pegasus 103 (F402-RR-401) by 600 lbf. Two new lift improvement devices (LID), longitudinal fences attached to the gun pods

The British Aerospace big-wing proposal was decidedly different from McAir's. The design featured wingtip missile rails, and the wing was to be made of conventional aluminum instead of composites. (British Aerospace)

The first YAV-8B was an AV-8A modified with the new wing and engine nozzles. The fuselage was unmodified, explaining the original low cockpit. A "normal" AV-8A flies chase. (McDonnell Douglas)

and a retractable lateral fence at the forward end of the pods, trapped a cushion of air beneath the fuselage and also helped prevent ingestion of recirculated gases from the exhaust nozzles. These added the equivalent of 1,200 pounds of extra lift, equating to an equal payload increase during VTO operations. The LIDs had been developed by Hawker much earlier but had never been used since it always seemed easier to increase the thrust of the Pegasus engine. It is interesting to note that the LIDs turned out so efficient that whereas pilots of earlier Harriers had to increase power as they approached the ground during vertical landings, Harrier II pilots have to cut power to dump lift.

Other airframe refinements included a 4° reduction in wing sweep to improve longitudinal stability at high angles of attack, while the aspect ratio was increased to achieve better cruise performance. The increased span and aspect ratio allowed the use of six underwing stores pylons, as well as allowing the roll-control thrusters to be positioned further outboard, thereby improving lateral control without increased engine bleed. To improve the turning radius during

The second full-scale development (FSD) AV-8B made its first flight on 7 April 1982. The LERX was added to this aircraft at the insistence of the RAF, which needed more maneuvering performance than the baseline McDonnell Douglas design was going to achieve. The aircraft is overall white with day-glo orange, royal blue, and gold trim. (Dennis R. Jenkins)

ground operations, the outrigger wheels were moved closer inboard, resulting in a reduction of wheel track from 22.18 to 16.99 feet. This would allow Harriers to operate from single-lane roadways for the first time.

The 28-foot wing was the largest single aircraft structure to be constructed of carbon-fiber at the time. The wing would be 400% stronger than a similar all-metal unit and saved 330 pounds. In order to make the wing, laminated sheets of carbon-fiber were heated to 500°F in an autoclave under 200 psi pressure. The new wing was fitted with large single-slotted, positive-circulation flaps which lowered to 61°, and added 7,600 pounds of lift during STO operations.

Formal US approval of the Harrier II development project (designated AV-8B) came on 27 July 1976, but the British were still not committed to the new aircraft. Budgetary problems caused by rampant inflation meant that no long-term decision on new aircraft would be made quickly. A no-longer flyable AV-8A (BuNo 158385) was modified with many of the AV-8B's features and tested in the 40x80-foot wind tunnel at NASA's Ames Research Center. Further development resulted in a new forward fuselage with a raised cockpit and bubble canopy, and a lengthened rear fuselage.

Two prototypes were converted from two AV-8As (BuNo 158394 and 158395) using the new wing, LIDs, and air intakes, but not the new forward fuselage and cockpit or the lengthened rear fuselage.

The AV-8B was the first Harrier to be put through an extensive set of spin tests. As a safety measure, a special parachute pack was installed on top of the tail stinger. If the pilot got into an unrecoverable spin, he could deploy the parachute, which was intended to put the aircraft into a recoverable attitude. Thankfully, the parachute did not have to be used during the test program. The unique double-row of pressure relief doors was only installed on the four FSD aircraft. The second row was deleted from production aircraft after the internal passageways were redesigned to accomplish much the same purpose. To aid in photography, the vertical stabilizer of this aircraft was painted day-glo orange. (Craig Kaston via the Mick Roth Collection)

The fuselage strakes of the AV-8B were significantly larger than on earlier Harriers. The "air dam" across the fuselage between the strakes was retractable and only deployed during hover when it was necessary to increase the ground effect. Photographed on 16 October 1988. (Keith Snyder via the Mick Roth Collection)

These aircraft were also fitted with new YF402-RR-404 engines that featured distinctive extended zero-scarf forward nozzles which were more fully enclosed and prevented the jet efflux from splaying outward (hence increasing ground effect). The first of these made its maiden flight on 9 November 1979 in St. Louis. The second made its first flight on 19 February 1979, but was lost on 15 November 1979 when the pilot ejected after his engine flamed-out.

The YAV-8Bs proved to have much greater drag than the earlier aircraft, even after modifications to the wing-root fairing, intake cowls, and inboard underwing pylons in an attempt to cure the problem. This made the aircraft significantly slower in most speed regimes. However, four months of extensive testing showed that the YAV-8Bs were meeting the Marine performance and payload requirements. Brief consideration was given to fitting the new wing and LIDs to existing AV-8As, but the option was eventually rejected when it was discovered that the empennage would need some serious structural reinforcement to cope with the new loads.

During this time, and expecting a long drawn-out battle in Congress to fund production AV-8Bs, the Marines opted to convert a limited number of AV-8As to the AV-8C configuration. The British government could still not make up its mind whether it wanted to participate in the Harrier II program, develop its own advanced Harrier version, or simply do nothing. Finally, as part of the FY79 defense bill, the US Navy authorized McAir to proceed with the AV-8B program, and established a requirement for 336 production aircraft to reequip the three AV-8A/C squadrons plus five A-4M squadrons.

By delaying their decision to join the Harrier II program, the British found themselves in the position of having to accept the capabilities of the AV-8B without recourse for serious changes. This was initially hard to accept since the operational requirements for the two countries were vastly different. However, the substantial savings that were achieved by allowing the Americans to largely foot the development bill, plus the lowered cost of acquiring aircraft from a large production run, eventually proved attractive to the RAF. A requirement for 60 RAF variants was eventually established for planning purposes.

Therefore, in August 1981 a second memorandum of understanding was signed between McAir and BAe, this time detailing the split of manufacturing responsibilities between the two companies. McAir would build approximately 60% of the airframe (as measured by man-hours) while BAe would build the other 40%, including the aft and

Unlike the RAF's GR5, the AV-8B did not have dedicated pylons for AIM-9 Sidewinders. This meant that the outer weapons pylons were generally used to carry Sidewinders. (David F. Brown via the Mick Roth Collection)

center fuselages, and rudders used on all Harrier IIs, and the single-piece composite horizontal stabilizers eventually used on RAF GR5s.

All aircraft supplied to the USMC and foreign nations are assembled by McDonnell Douglas in St. Louis, while all RAF aircraft are assembled by British Aerospace in Dunsfold. A similar agreement was signed between Rolls-Royce and Pratt & Whitney under which the British company would be responsible for 75% of the engine's manufacture, and all final assembly. These agreements did not cover mission-unique equipment such as avionics simply because it was expected this would be nearly completely different between the two nations.

Six complete rear fuselage assemblies were delivered from Kingston to McAir for use in the first four Full-Scale Development (FSD) AV-8Bs (BuNo 161396/399) and two static test airframes. The FSD aircraft were equipped with the new McAir-designed forward fuselage, which was based loosely on the unit fitted to the Sea Harrier. The forward fuselage was of composite construction, and weighed approximately 25% less than the AV-8A's forward fuselage. The pilot's seat and cockpit floor were raised, allowing more room for systems and equipment. A bubble canopy feature much improved rearward vision, and a new single-piece wrap-around windscreen eliminated the forward windshield frames that obstructed the pilot's vision on earlier aircraft. An extended rear fuselage compensated for the change in center of gravity due to the new wing and a redesigned horizontal stabilizer was used.

The first FSD aircraft was flown on 5 November 1981 piloted by Charles A. Plummer. The next two aircraft followed on 7 and 10 April 1982, with the fourth and final flying on 4 June 1983. The first two aircraft featured large nose pitot booms, while the third and fourth introduced a wing leading edge root extension (LERX) necessary to meet RAF turn performance requirements. This LERX was a "70% version" for the unit originally designed by Hawker, although a "100% version" would make its appearance in later production aircraft.

The final FSD aircraft was completed to near-production standards and was used for weapons trials, including test of a new 25-mm cannon. The original Harriers had carried two Aden 30-mm cannon pods, one replacing each of the under-fuselage strakes. Although excellent weapons, they proved very difficult to support in US service since they used unique (to the Americans) ammunition and components. The AV-8B would carry a single GAU-12/U multibarrel 25-mm cannon mounted in the left pod, with 300 rounds of ammunition contained in the right pod, using a "bridge" along the fuselage to transfer the ammunition. The new cannon used engine bleed air to drive it to a firing rate of 3,600 rounds per minute. The ammunition was the same as used by various 25-mm cannon fitted to Marine armor vehicles, greatly simplifying logistics.

All AV-8Bs would be capable of in-flight refueling, and tests were conducted using the second FSD aircraft and a Marines KC-130F tanker. However, even without refueling the AV-8B had impressive range. In March 1983 Charles Plummer flew one of the FSD aircraft equipped with seven 500-pound bombs and full internal fuel 422 miles to a target at extreme low level, dropped the bombs, then returned at high altitude. He had taken off using only 800 feet of runway, and still had 1,800 pounds of fuel remaining after his vertical landing. Testing revealed that maximum STO take-off weight was 26% greater than the AV-8A (31,000 versus 24,600 pounds). This was demonstrated when Plummer flew a CAS profile with twelve 500-pound bombs, taking off after a 1,200-foot roll and attacking a target 185 miles away, returning for a vertical landing with 600 pounds of fuel remaining. On missions of the same length, the AV-8B could carry twice the ordnance of the AV-8A.

The AV-8B featured the same Hughes ASB-19(V)-2 Angle Rate Bombing System (ARBS) that was well proven in the A-4M Skyhawk. It was optimized for dive attack profiles and was linked to the AYK-14 standard mission computer and SU-128/A HUD. The cockpit was very different from the analog instrumentation used in earlier Harriers. Large multifunction displays and HOTAS (hands-on-throttle-and-stick) almost identical to the ones developed for the F/A-18 were fitted. The ARBS uses collimated TV and laser spot trackers located in the extreme nose to provide up to a 6x magnified view on one of the MFDs.

A sophisticated three-axis Sperry stability augmentation and attitude hold system (SAAHS) minimizes pilot workload during hover and transition to wingborne flight. The SAAHS is technically capable of allowing "hands-off" vertical landings to be made, as demonstrated by Bill Lowe in February 1983. The

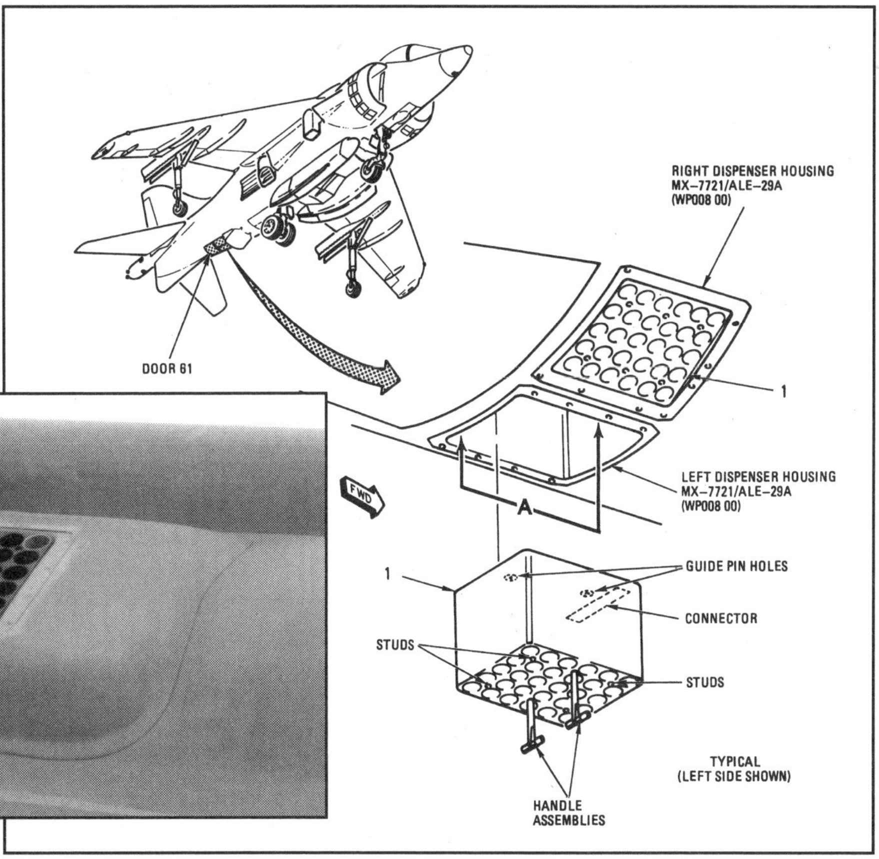

All AV-8Bs are equipped with two chaff/flare dispensers located under the fuselage just behind the speed-brake. (US Marine Corps)

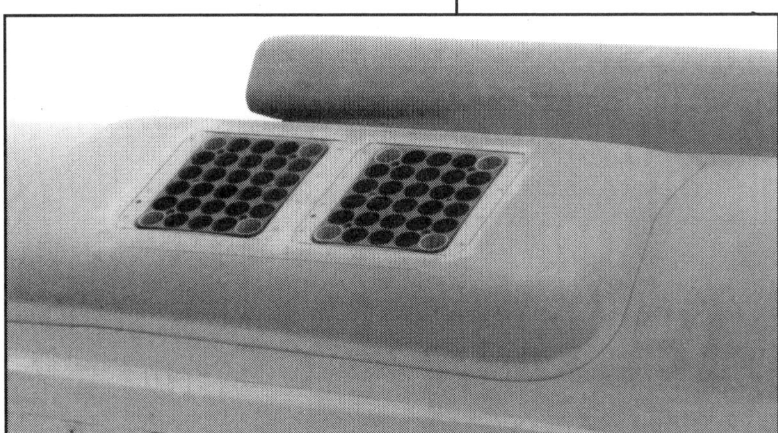

The AV-8B(NA) and AV-8B+ are equipped with two additional ALE-39 dispensers on top of either side of the fuselage. (Chris M. Reed)

AV-8B featured ALR-182 radios with KY-28 secure voice systems, APX-100 IFF transponder, ALE-39 expendable countermeasures dispensers, ALR-67 radar warning receiver, and provisions carrying an ALQ-164(V) deceptive countermeasures pod on the centerline station. This pod is capable of countering mid- and high-band pulse and continuos-wave (CW) threats. Eventually an AAR-44(V) UV/IR missile approach warning system (MAWS) was also fitted into the tail stinger of the AV-8B to warn of missiles approaching from the rear.

The first pilot-production AV-8B (BuNo 161573) took to the air on 29 August 1983, followed by two more by the end of the year. The major external change from the prototypes was the deletion of the second row of blow-in doors around the air intakes, made possible by a slight redesign of the internal passageways. This aircraft was officially turned over to VMA(T)-203 at MCAS Cherry Point on 12 January 1984.

The Marines believed they could get by, at least for a while, without a two-seat derivative of the new Harrier II. But the new aircraft exhibited decidedly different landing characteristics than the AV-8A, and the transition between the analog cockpit used in the TAV-8A and the new AV-8B was proving difficult for new pilots. As a result, at the beginning of 1984, eight of the AV-8Bs were cancelled and the money saved diverted to the development of a two-seat trainer for the new aircraft. Twenty-eight two-seat TAV-8Bs were procured for the USMC between 1985 and 1989.

The first TAV-8B (the 65th Harrier II) flew on 21 October 1986, with deliveries to the VMA(T)-203 beginning on 24 July 1987. A completely new tandem-seat nose was used that stretched the fuselage 3 feet 11 inches. The vertical stabilizer was made 1 foot 5 inches higher and given an increased chord to com-

This formation flight gives a good size comparison against the Northrop F-5E Tiger II. (Mick Roth Collection)

The TAV-8B, unlike the earlier two-seaters, is not a mission capable aircraft, and only has two under-wing weapons pylons. The vertical stabilizer chord has been increased to compensate for the larger canopy area. (Craig Kaston via the Mick Roth Collection)

pensate for a slight loss in directional stability. Unlike the TAV-8A, the new two-seater does not have an extended tail stinger. Also unlike the original two-seater, the TAV-8B was not designed to be used operationally as an attack aircraft. Therefore, it is only equipped with two external weapons stations, intended primarily to carry fuel tanks or training weapons.

The first 12 production AV-8Bs were powered by F402-RR-404A turbofans that were generally similar to the -401 that had powered the AV-8A/C. By late 1984 one of the FSD aircraft was flying with a new -406 engine that used a modified turbine section that ran 20-30°F cooler at any given engine speed, significantly increasing reliability and the mean time between overhauls (up to 1,000 hours). The engine passed a quick service evaluation and was introduced on the production line in mid-1985. The addition of digital fuel controls resulted in the -406A engine, roughly equivalent to the RAF's Pegasus Mk 105 engine. The prototype TAV-8B (BuNo 162747) was the first aircraft to use the -406A. Beginning in 1990 all aircraft were completed with the -408 version of the engine that increased thrust to 24,500 lbf – the same level originally contemplated for the PCB-equipped Pegasus 15.

It should be noted that the thrust figures quoted for the Pegasus, like most jet engines, are hopelessly optimistic. The F402-RR-404 (Pegasus 11) has a nominal short (brief takeoff or hover) rating with water injection of 21,500 lbf, or 20,500 lbf without the water injection. This rating is available for a maximum of 15 seconds per flight. A normal maximum rating of 19,500 lbf is available for up to 150 seconds per flight, normally during VTO operations. In conventional wingborne flight the engine is limited to 16,750 lbf for up to 15 minutes, with a maximum continuous rating of 13,500 lbf.

The first operational squadron to receive the new aircraft was VMA-331, which traded their McDonnell Douglas A-4 Skyhawks. The squadron was recommissioned on 30 January 1985, and was declared fully operational in September 1986. The AV-8B was quick to demonstrate that it met the USMC maintenance requirements of 15.9 MMH/FH (maintenance manhours per flight hour), a 1.9-hour mean time to repair, and a 75% mission readiness rate.

The AV-8B was the first Marine combat aircraft to arrive for Operation Desert Shield/Storm in 1991. In all, the Marines deployed 86 AV-8Bs which flew 3,380 missions totaling 4,112 combat hours. More than six million pounds of ordnance were dropped by the Harriers during 42 days of operations. During the war the Harriers operated from a semi-unprepared airfield owned by an oil company at Tanajib in Saudi Arabia, and off the USS *Nassau* (LPH-6). After the war, General Schwarzkopf specifically named the AV-8B as being one of the seven weapons systems that had made a significant contribution to a quick and decisive victory. The only other aircraft named were the F-117 and AH-64.

This April 1991 photograph shows the significant variations that existed in the standard camouflage pattern. Carefully compare the vertical stabilizer and forward fuselage – it is not simply a "reverse" paint scheme. (David F. Brown via the Mick Roth Collection)

The AV-8B can carry a retractable refueling probe on the left air intake. The AV-8A had featured a non-retractable probe in the same location. The larger external fuel tanks are noteworthy. Photographed on 18 May 1991. (Keith Snyder via the Mick Roth Collection)

An AV-8B(NA) hovers over the deck of the USS Pelileu *(LHA-5) on 20 June 1997.* (US Navy)

Five of the eight available Marine Harrier squadrons were deployed, in whole or in part, to the Gulf during the war. The aircraft flew mainly against targets in Kuwait using CBU-20 anti-armor cluster bomb, napalm, Mk 82 500-pound bombs, Mk 83 1,000-pound bombs, and AGM-65E semiactive laser-homing Maverick missiles. Early in the war AIM-9s were carried for self defense, but with the virtual disappearance of the Iraqi air force, these were later omitted. One aircraft in each strike package usually carried a Sanders ALQ-164 ECM pod.

Originally the AV-8B's ram air intake (at the base of the vertical stabilizer) was relatively short (as seen here). This would change as the aircraft went through later modifications, and the final intake extends forward almost to the trailing edge of the wing. (Michael Grove via the Mick Roth Collection)

The uncluttered nose of the AV-8B (above) would soon change. (Mick Roth)

The Night Attack Harrier – designated AV-8B(NA), even if the designation system does not allow such – added a small bump on top of the nose to contain a FLIR. Together with NVGs, this gave the Harrier a credible night attack capability. (Kenneth Kula via the Mick Roth Collection)

A total of five AV-8Bs were lost to enemy action during the war. A later analysis concluded that four of these could possibly have been saved if the MAWS missile detection system had been operational. The Harrier is particularly vulnerable to small heat-seeking missiles. During the Gulf war, several fighters (F-15 and F/A-18) limped home after a missile hit the extreme rear of their exhaust nozzles. On the Harrier, missiles tend to home onto a spot between the four nozzles, usually resulting in the aircraft being completely destroyed.

Night Attack

In late 1985 the Navy awarded McDonnell Douglas a $2.1 million design definition contract for a night attack variant of the AV-8B. Originally known as the AV-8D, politics conspired to call the new aircraft AV-8B(NA). A test was conducted using a GEC-Marconi FLIR and GEC Cats Eyes night-vision goggles (NVG) on a TA-7C at China Lake. The test was generally considered successful, and the two components formed the nucleus of the new AV-8B(NA) modification.

In 1989 GEC received a $16 million contract for FLIRs to be mounted in a small fairing immediately in front of the pilot on top of the nose. This fairing was larger and of different contours than the similar fairing installed on later RAF GR7s. A new wide field-of-view HUD presented the FLIR information, and third-generation GEC Cats Eyes III NVGs were also provided.

Unlike later RAF aircraft, the AV-8B(NA) was not equipped with an automatic method of jettisoning the NVGs from the pilot's helmet during ejection. This has led to some restrictions about when the pilot can wear the NVGs since ejecting with the unit in place can seriously injure the pilot. The aircraft also featured new color MFDs, and a CRT-based digital map display that retrieved its data from Hamilton Standard ASQ-194 CD-ROM data storage set.

The AV-8B(NA) was equipped with the F402-RR-408 (Pegasus 11-61) engine which features improved fan aerodynamics and single crystal turbine blades, along with other smaller improvements. The hot inspection cycle has increased to 1,000 hours (from 500 for the -406) and total life-cycle costs have been reduced by over 40 percent.

Another visual change to the Night Attack Harrier II is the addition of four Tracor ALE-39 expendable countermeasures dispensers above the rear fuselage. Combined with the two ALE-39 dispensers located on the bottom of the fuselage, this gives the AV-8B(NA) a total of 180 flares or chaff cartridges. The RAF-style "100%" LERX was included, and the ram air intake at the base of the vertical stabilizer was lengthened, almost to the trailing edge of the wing.

The 87th single-seat AV-8B (BuNo 162966) was selected as the Night Attack prototype, first flying on 26 June 1987. The first production AV-8B(NA) was the 167th aircraft (BuNo 163853), and all subsequent aircraft would be built to this standard until the advent of the Harrier II Plus (AV-8B+). VMA-214 was the first unit to receive the Night Attack variant beginning in September 1989, and all four Yuma-based units were so equipped by September 1992. The new engine initially displayed some engine problems that were cured by switching from titanium fan blades to common steel ones, and increasing the fan-blade clearance slightly.

Harrier II Plus

During 1987 McAir accomplished preliminary design work on an aircraft tentatively called the AV-8E. This design used a more powerful variant of the Pegasus engine, larger LERX extensions, and featured a radar offering both air-to-air and air-to-ground modes. Funding limitations kept the Marines from pursuing this aircraft.

In 1988 BAe and McAir announced they were developing, as a private venture, a radar-equipped version of the Harrier II. Except for the Sea Harriers, all previous Harriers had not been equipped with radar. In September 1990, the US, UK, Spain, and Italy signed a memorandum of understanding to jointly develop and produce the new variant. Alenia and CASA each gained a 15% share of the program.

The radar selected for use in the aircraft was, surprisingly, not a version of the GEC-Marconi Blue Vixen used in the FA2 Sea Harrier. instead, the Hughes APG-65 pulse-Doppler

Three external changes mark the AV-8B(NA): the bump on top of the nose that contains the FLIR; a pair of ALE-39 expendable countermeasures dispensers on each side of the upper rear fuselage; and an extended ram air intake at the base of the vertical stabilizer. Extending the ram air intake was necessary to ensure it did not ingest combustion products from the flare/chaff dispenser. The aircraft is carrying a ALQ-164 DECM pod on the centerline station. (Michael Grove via the Mick Roth Collection)

All big-wing Harriers can carry a retractable in-flight refueling probe on the left air intake. This allows them to refuel from any tanker equipped with a standard probe-and-drogue setup, which is used by the US Navy and many foreign air forces. The latest US Air Force tankers (KC-10) also carry the appropriate equipment. (David F. Brown via the Mick Roth Collection)

Some AV-8B(NA)s were painted a light grey with medium grey subdued markings. This improved the appearance of the aircraft substantially. Unusually, this aircraft does not have any underwing pylons attached. (Mick Roth)

radar used in the F/A-18 was chosen, albeit with a smaller antenna in order to fit into the Harrier's modified nose. This decision brought mixed reviews. The primary rationale for choosing the radar was that the F/A-18 was being upgraded to the newer APG-73, and a large quantity of used, but still very serviceable, APG-65s were available for very little cost.

However, the APG-65 was not the most capable unit, and it had not been integrated with the new AIM-120 AMRAAM. In contrast, the Blue Vixen was integrated with the AIM-120, and in fact had been considered as a candidate for replacing the APG-65 in the F/A-18. But the fact remained that it cost a great deal more, and budgets were tight.

The addition of the radar brought with it a 1,000-pound increase in gross weight, but this did not significantly decrease the performance of the aircraft, all of which would be equipped with the -408 engine. Perhaps the most significant external change, other than the new nose profile, was the larger RAF-style "100%" LERX, although it has a slightly different shape. The Harrier Plus included all of the night attack equipment from the AV-8B(NA), although the FLIR faring on top of the nose was reshaped slightly.

The second FSD AV-8B was modified to serve as an aerodynamic prototype for the Harrier Plus, and was frequently shown carrying AIM-120s on the wing pylons. The aircraft also was shown carrying AGM-84 Harpoons, although full integration of this missile is far from certain. A $20 million contract from the Naval Air Systems Command allowed McAir to modify the 205th single seater (BuNo 164129) to serve as a full-up prototype. The modified aircraft first flew on 22 September 1992 with Jackie Jackson at the controls.

Originally the Marines ordered 27 new-build aircraft, designated AV-8B+ (yet another designation that does not conform to the standard). The intent was to order a further batch of 48 new-builds, and also to modify all remaining AV-8Bs to the new configuration. Budgetary considerations resulted in the second order not being placed. Current plans are still to convert all remaining AV-8Bs as money permits, although only 73 conversions have been funded as of mid-1998. The new-build aircraft were delivered to three squadrons at MCAS Cherry Point.

The conversion of older AV-8Bs to the new configuration is not simple, and costs roughly 66% as much as a new aircraft. A completely new forward fuselage, -408 engine, radar, FLIR, and cockpit displays are added to the basic center and rear fuselage from the original aircraft. The wing, landing gear, and weapons pylons receive rework.

A number of other upgrades are currently being funded to increase

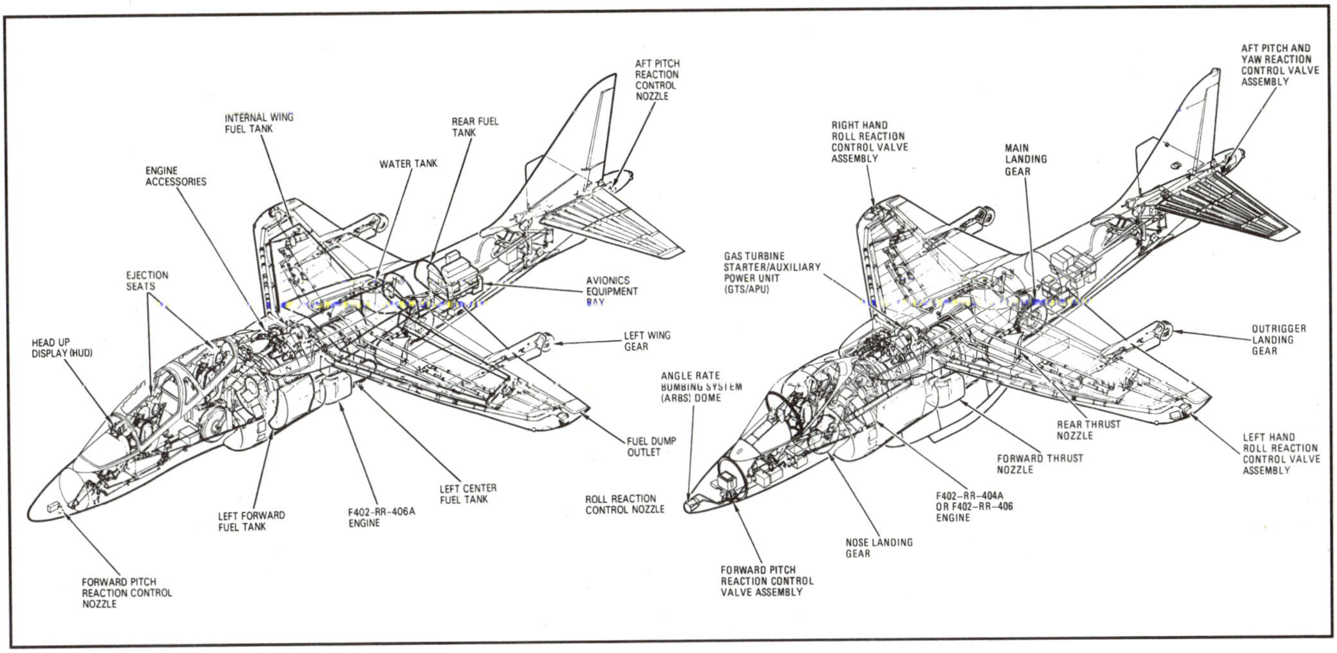

The general arrangement of the TAV-8B (left) and AV-8B (right) shows the location of major equipment. Of note are the fuel tanks that flank the front of the engine, appearing very vulnerable to ground fire. (US Marine Corps)

the capability and reliability of the AV-8B. These include the open systems core avionics requirements (OSCAR) which makes it possible to integrate commercial-off-the-shelf (COTS) avionics systems into the aircraft. At the same time the aircraft are being equipped with in integrated GPS receiver, ARC-210 radios, ALE-47 expendable countermeasures dispensers, and a digital incident recorder.

As the AV-8B continues to gain new capabilities, it it interesting to note how this affects its unit cost. The last "plain" AV-8Bs cost $14.9 million each in FY87; the AV-8B(NA) cost $17.6 million each in FY90; while the FY92 AV-8B+ cost $31.9 million each. These costs include the engine and avionics.

One concept unlikely to see the light of day is the AV-8SX demonstrator. Yet another attempt to build a supersonic Harrier, this aircraft would use a Pegasus 11F-35 with PCB along with a modified Harrier II airframe.

Although it is difficult to see, this AV-8B+ from VX-9 has been fitted with the "100%" LERX. These have a slightly straighter leading edge, instead of the gentle curve of the "75%" units. (Mick Roth)

Originally, the AV-8B+ update was to include the eight-pylon RAF-style wing, but this was not pursued for cost reasons. (Terry Panopalis)

The ALE-39 countermeasures dispensers are not installed on this aircraft, although the "bumps" where they are mounted are evident. The dispensers themselves are normally only carried when the aircraft is on a training mission, or almost any time the aircraft is outside the United States. (Michael Grove via the Mick Roth Collection)

The AV-8B+ (shown here at China Lake in 1993) added a formidable APG-65 radar capability to the Harrier. This was a major step for the Marines who have always maintained that the AV-8 should be as simple as possible. (Craig Kaston via the Mick Roth Collection)

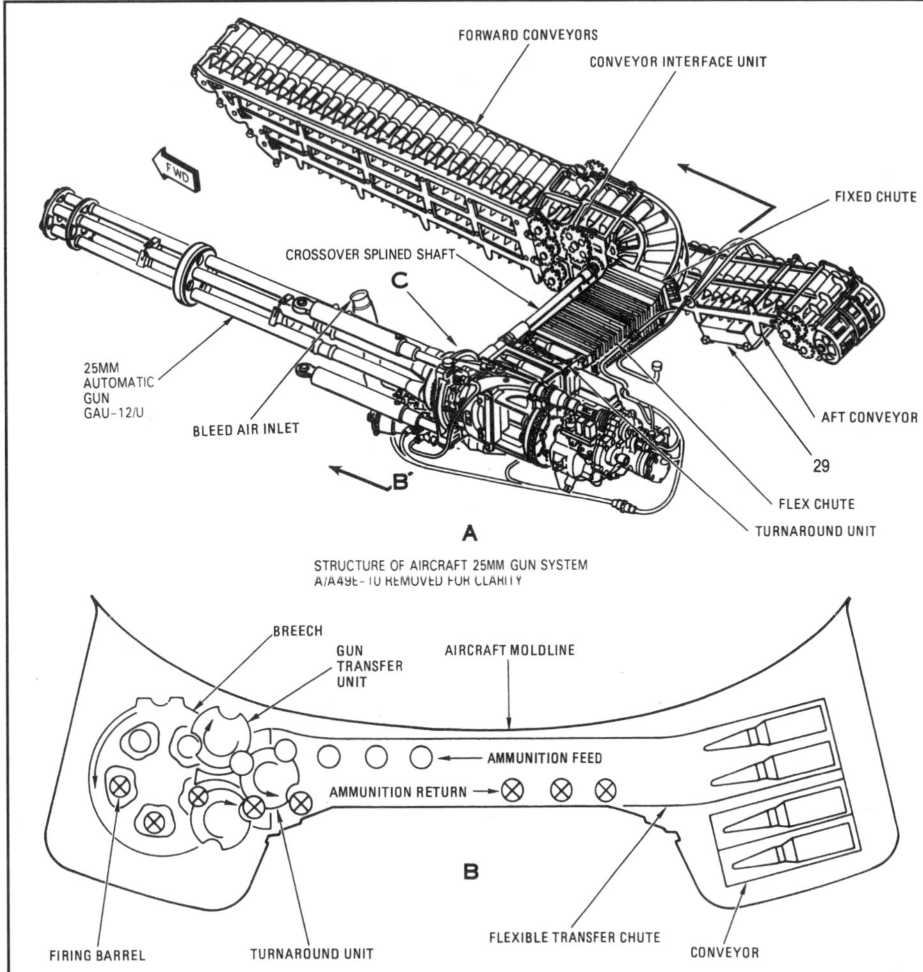

Figure 1. Aircraft 25mm Gun System A/A49E-10 Component Locator (Sheet 9)

To simplify their logistics, the Marines chose a 25-mm cannon that was already in their inventory for use on several ground vehicles. Although the actual cannon differs in many details, the ammunition used is the same. The packaging of the pods is interesting, with the cannon itself residing in one pod, while all of its ammunition resides in the opposite pod. A flexible chute between the pods transfers the ammunition. The RAF elected to stay with a pair of Aden cannon, for many of the same reasons (simplified logistics). (US Marine Corps)

The large "zero -scarf" front nozzles are vastly different from the nozzles used on first generation aircraft.These nozzles are more fully enclosed and prevent the jet efflux from splaying outward. (Boeing)

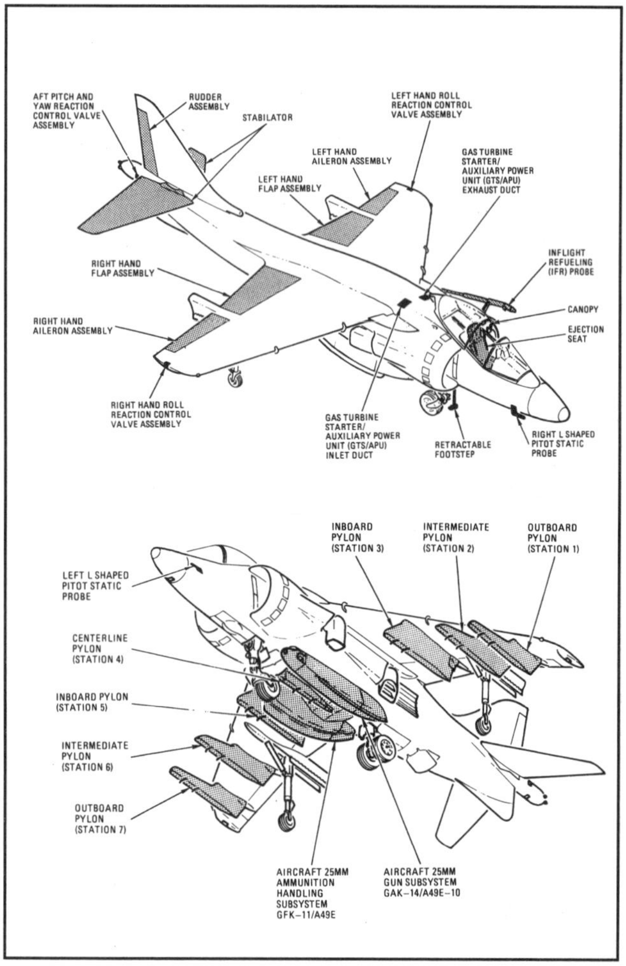

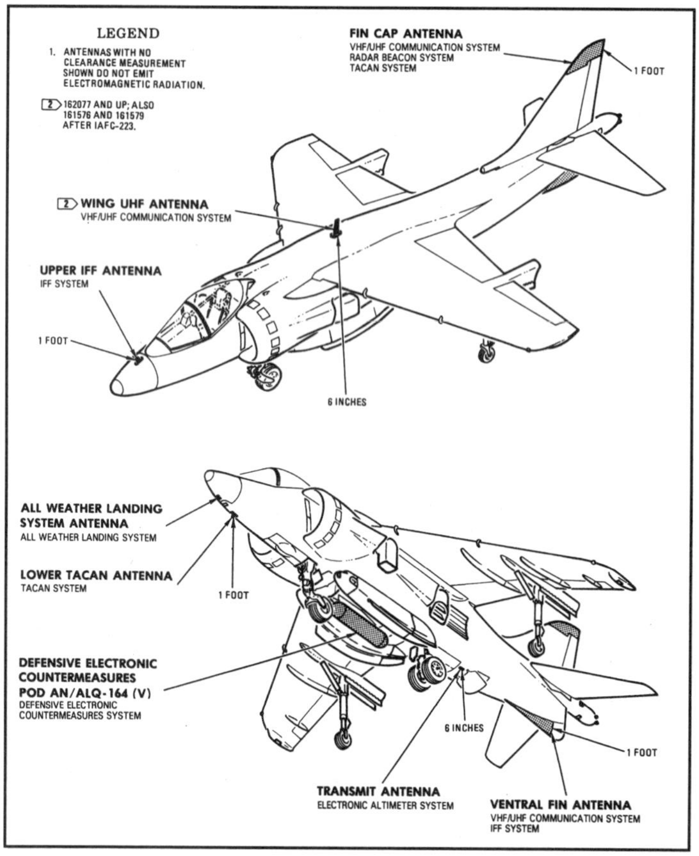

Various antennas (above) and danger areas (left) around the AV-8B. (US Marine Corps)

Import 9 Export

The British, Spanish, and Italians Buy Big-Wing Harriers

After pulling out of the AV-16/P.1184 project, it seemed Britain would opt not to procure any second generation V/STOL aircraft. For a while the idea of retrofitting the new larger wing to the existing GR3 was considered. The Kingston-developed wing was decidedly different than the one being used on the AV-8B. Although perhaps even more advanced aerodynamically, the BAe wing was of conventional aluminum alloy construction, and became known as the "tin wing." One of the major benefits of the BAe wing was a lower drag coefficient at high speed, potentially solving some of the drag problems encountered on the AV-8B. The downside to the wing was its weight, a full 400 pounds more than the McAir wing. The Kingston wing also included wing-tip rails for Sidewinder-type missiles.

However, the RAF believed in flying their Harriers a lot, and by 1988 many of the aircraft had amassed over 4,000 flight hours. Installing new wings on fuselages that had already used the majority of the useful life was not deemed cost effective. As an alternative, BAe proposed installing the new wing on new-production aircraft that used the basic Sea Harrier fuselage (to gain the raised cockpit). This aircraft, informally known as the GR5(K) also incorporated the AV-8B's revised intakes, zero scarf forward nozzles, and most of the LIDs. Unfortunately the RAF's requirement for any GR3 replacement involved only 60 aircraft to replace the Harriers based in Germany. The UK-based squadrons would continue to operate upgraded GR3s. A 60-aircraft order was insufficient to justify the development of the new aircraft, particularly when the competing AV-8B already had over 300 potential orders.

The British government was fully committed to a couple of rather expensive defense projects, namely the Trident missile submarine and the Tornado strike aircraft. It simply could not afford to support another major development program. In the meantime, the American AV-8B had been demonstrating impressive results. Finally, in August 1981 the British decided to participate in the AV-8B program, and a memo-

Harrier GR7 shows the "75%" LERX initially used as a compromise in an ongoing dispute with the USMC. RAF Harriers are painted an overall dark green with small subdued markings. (Nick Challoner)

randum of understanding was signed, effectively killing the BAe big-wing Harrier. When the RAF finally joined the Harrier II program, it expected each of the 60 production G.R. Mark 5 to cost approximately £9 million ($14.4 million) each. By the time the first aircraft flew, this had increased to £11 million ($17.6 million), largely due to inflation and changes in equipment.

Although the RAF decided to procure the AV-8B airframe, this did not apply to the American avionics. The Marines wanted a simple aircraft that was easily maintainable in the field under very austere conditions. They were more than willing to settle for a very basic attack system since the missions they were flying were essentially close air support, and required very little in the way of sophisticated aiming equipment. On the other hand the RAF needed a versatile strike aircraft capable of operating against sophisticated Warsaw Pact defenses.

A large Smiths Industries SU-128/A heads-up display is located in the center of the instrument panel, flanked by the Hughes ASB-19 Angle Rate Bombing Set to the left, and a Ferranti moving map display to the right. Many pilots believed that the LRMTS used on the GR3 was more capable than the ARBS. A Marconi ZEUS radar warning receiver and electronic countermeasures system is capable of providing 360° warning and jamming. The system can identify over 1,000 separate types of emitters, and select the correct countermeasures for each of them. In the extreme tail is a Plessey pulse-Doppler radar missile approach warning system (MAWS) which can sense the approach of missiles and respond by launching chaff and/or flares from the ALE-40 expendable countermeasures dispenser under the rear fuselage. Originally a CelsiusTech BOL chaff dispenser was to be located in each outrigger fairing, but these were never installed.

The GR5 was originally scheduled to use a Ferranti FIN.1075 inertial navigation system, but after repeated delays during development, the first batch of GR5s was delivered with the same Litton ASN-130 INS units used on the AV-8B. Continued delays forced the RAF to procure 32 additional ASN-130s to equip the first squadron of GR5s in Germany. Eventually the Ferranti system passed its service tests and all earlier aircraft were retrofitted. Beginning in 1992 an integrated GPS receiver was added to the INS, creating the FIN.1075G. A Cossor IFF.4760 transponder replaced the AV-8B's APX-100 unit, while a GEC AD.3500 radio with secure voice replaced the equivalent American systems.

During 1980 Rolls-Royce (who had acquired Bristol) began the development of a digital engine control system (DECS) for the Pegasus engine with assistance from Smiths Industries and Dowty. The DECS was fitted to a Pegasus Mk 103 in a GR3 (XV277) which made its first

A GR5 from the 233rd OCU during 1992. The extra pylon on the wing gave the RAF decidedly more weapons flexibility than the USMC, at only a small penalty in extra weight and drag. The Aden gun pods are smaller than the USMC GAU-12 pods, although they carry less ammunition as a result. (Kev Darling/BBA)

The BL755 cluster bomb was intended to be the Harrier II's primary weapon. Here a GR7 carries a BL755 on the two outboard pylons, an AIM-9 on the outrigger pylon, and a fuel tank inboard. Another BL755 is mounted on the centerline pylon between the two gun pods. (Nick Challoner)

flight on 11 March 1982. The next five years were spent refining the software for the system. The end result was an engine capable of smooth operation at all throttle positions. One of the problems that had been encountered by all earlier Pegasus engines was their tendency to stagnate at 92% thrust. The incorporation of DECS as well as other detailed improvements led to the production of the Mk 105 engine for the GR5, as well as the F402-RR-406A for the AV-8B.

Other differences between the AV-8B and GR5 included the addition of an additional weapons station forward of the outrigger pylons. These pylons can only carry the AIM-9 Sidewinder, but this frees the other pylons for offensive weapons. A ceramic coating covers the outrigger fairings to protect them from the missile exhaust. The outer pylon can also carry a Sidewinder, just like the AV-8B. All of the underwing pylons were slightly redesigned to allow them to carry the BL755 cluster bomb, which was the primary weapon envisioned for the GR5. The BL755 is heavier and bulkier than the Mk 82 iron bomb, around which the USMC AV-8B was designed. The high-speed low-level missions envisioned for the GR5 dictated the ability to withstand a one-pound bird strike at 690 mph. To meet this the wing leading edge skin was increased in thickness and the intake lips and windscreen were also strengthened.

The RAF elected not to use the new 25-mm cannon adapted by the USMC. Instead a pair of gas-operated Aden 25-mm cannon are carried. Compared with the older 30-mm cannon carried by the GR3, the new cannon delivers three times the kinetic energy, mainly through a higher muzzle velocity and higher rate of fire. Each pod carries 100 rounds for its cannon. The British cannon are slightly smaller than the American system, allowing more streamlined pods. The pair of cannon and their ammunition also weigh 200 pounds less than the GAU-12, but the American cannon has an additional 100 rounds of ammunition (300 total). Interestingly, as late as mid-1998 it is uncertain if any of the cannon have ever been carried by RAF Harrier IIs. The pods are routinely seen fitted to aircraft, but they invariably do not contain the cannon, whose barrel protrudes some distance from the front of the pod (unlike the GR3 and AV-8B pods). Some reports indicate that the right pod usually contains a GEC-Marconi ARI 23333 Zeus RF spectrum ECM system.

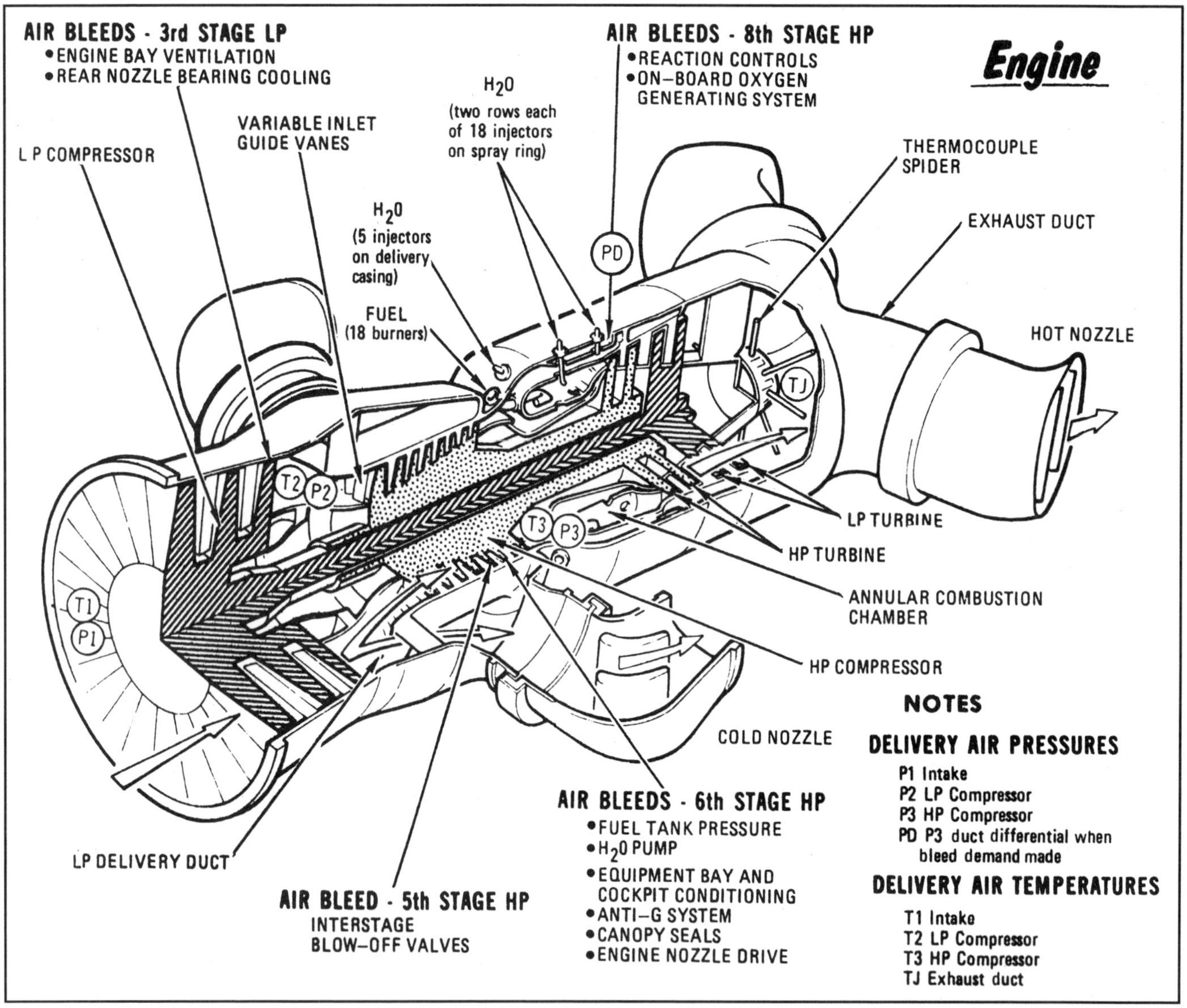

The Pegasus Mk 105 (F402-RR-406A in US service) features digital fuel control and greater reliability than earlier Harrier engines. (US Marine Corps)

All of the RAF's Harriers had always had a limited reconnaissance capability using their F.95 oblique camera, even if it proved to be less than ideal. No. 4 Squadron in Germany also carried an external reconnaissance pod on the centerline station. On the GR5, a small fairing beneath the nose was intended to house a miniature infrared linescan system (MIRLS) adapted from the system designed for the Tornado GR1A. This synthetic aperture reconnaissance system could scan from horizon to horizon. The intent was for all GR5s to be capable of carrying the equipment, but sufficient systems would only be procured to equip the squadrons based in Germany. In any event, funding problems prevented this system from being deployed and all GR5s flew with ballast installed in the fairings. A small number of GR7s were modified to carry the pods that had been carried on the GR3, although this proved to be remarkably expensive, taking an average of 650 man-hours per aircraft to accomplish. Eventually new reconnaissance pods were procured for the GR7.

The first Development Batch GR5 (ZD318) used the 20th set of Harrier II components. The aircraft was first flown at Dunsfold on 30 April 1985 with Mike Snelling at the controls. The second DB aircraft followed on 31 July. The official turnover of the first operational GR5 (ZD324) took place on 1 July 1987 at Wittering. Prior to the service introduction of the GR5, several GR3 pilots from No. 233 OCU

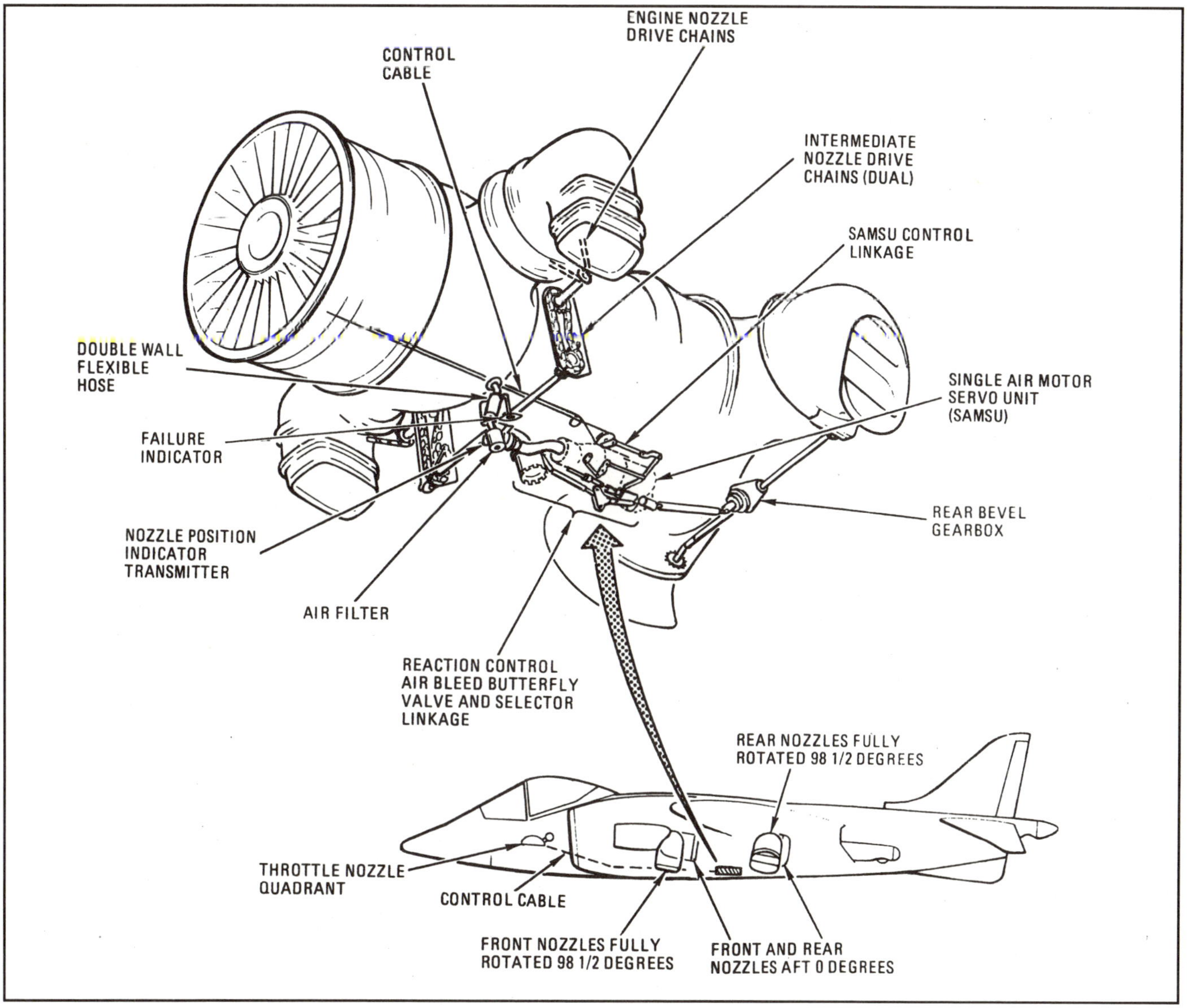

It is interesting to note that the nozzles can be rotated 8.5° to the front, (98.5° from the "zero" setting), allowing the Harrier to back up at a slow rate of speed. (US Marine Corps)

were dispatched to MCAS Cherry Point to gain experience flying the AV-8B. No. 1 (Fighter) Squadron began receiving GR5s on 28 November 1988, finally retiring their last GR3 on 31 March 1989. Among No. 1 Squadron's routine commitments were short-term deployments aboard the RN's Invincible-class carriers. As with previous Harriers, the GR5s were intended to serve in Germany, and No. 3 Squadron at Gütersloh received its first aircraft (ZD401) on 17 March 1989. No. 4 Squadron managed to skip the GR5 altogether, finally trading their old GR3s for new GR7s beginning on 7 December 1990.

On 1 November 1985 BAe test pilot Taylor Scott was killed in a freak accident involving the Martin-Baker Mk 12 ejection seat. Somehow the drogue chute accidentally deployed and fragmented the canopy, dragging Scott from the cockpit. Scott was killed, and the aircraft (ZD325) eventually crashed after it ran out of fuel. All Harrier GR5s were grounded pending the outcome of the accident investigation and subsequent modifications to ensure the problem could not happen again. The problem did not affect the AV-8B since the Americans continued to use Stencel ejection seats.

On 19 March 1988 the RAF ordered a further 34 GR5s, bringing the total to 94 (not counting the two DB aircraft). The first 41 were delivered as GR5s, while the next 21 were delivered as GR5As which incorporated wiring provisions for the FLIR being developed for the GR7. The last 34 aircraft were delivered as Harrier G.R. Mark 7s.

A single GR5 (ZD402) was modified to serve as a testbed for the Pegasus 11-61 engine. This engine features improved fan aerodynamics and single-crystal turbine blades and was based directly on experience gained during the XG-15 engine technology demonstration program. The gloss-black aircraft was used to set a series of Class-H time-to-climb records on 9 June 1989. Rolls-Royce chief test pilot Andy Sephton set records of 36.38 and 81.00 seconds, respectively, to 3,000 and 9,000 meters (9,842 and 29,527 feet). BAe test pilot Heinz Frick used the same aircraft to reach 6,000 and 12,000 meters (19,685 and 39,370 feet) in 55.38 and 126.63 seconds, respectively.

G.R. Mark 7

Like US Marine Corps, the RAF developed a FLIR system that allows an expanded role at night. A pair of Harrier T4As, along with a Tornado, Buccaneer , and a Jaguar, served as test aircraft for various nigh attack equipment under the Nightbird program. The Harriers featured cockpits modified to be compatible with night-vision goggles (NVG), and an experimental FLIR was installed in the LRMTS nose cones together with a new Ferranti 4510 raster HUD. The Nightbird aircraft provided a wealth of information regarding the advantages and limitations of NVG and FLIRs.

Based on the results from the Nightbird project, the RAF's GR7 received a new GEC-Marconi color map display and revised software for its Computing Devices ACCS 2000 (a license-built version of the US Navy/Unisys AYK-14) mission computer. An off-the-shelf GEC FLIR is mounted in a fairing in front of the cockpit on top of the extended (by nine inches) nose, as close as possible to the pilot's normal line of sight. The image from the FLIR can be presented on either the new wide-angle HUD or, more usually, a right MFD. A set of GEC-Ferranti NVGs complement the fixed FLIR, allowing the pilot to look around outside the cockpit. Because NVGs adversely affect a pilot's ability to eject from the aircraft (they "catch" the airflow and jar the pilot's head), Martin-Baker developed a compressed air system that jettisons the NVGs at the beginning of the ejection process, just in case.

There was little to distinguish GR7s from the earlier GR5s except for the modified nose profile and the incorporation of the 100% LERX. Interestingly, there have been two different 100% LERX designs, although only the one shown above seems to have been fitted to RAF Harriers. A second design, one with a mostly straight leading edge instead of curved, has been observed on various AV-8B+ aircraft. (Joris van Boven/Sentry Aviation News)

Beginning with the 17th new-build GR7 (the 77th Kingston-built Harrier II), a new "100%" LERX was fitted. This 7.5 sq ft LERX curved down towards its leading edge and was deeper in cross-section than the 75% LERX used on earlier aircraft. The larger LERX provided a substantial increase in instantaneous turn rate, and was subsequently retrofitted onto most earlier RAF aircraft and many AV-8Bs.

The RAF Harrier II had its combat initiation over Bosnia on 30 August 1995 when GR7s attacked an ammunition dump at Lukavica and an weapons factory at Vugosca. During June 1994 carrier trials were undertaken with a GR7, eventually clearing the RAF aircraft to operate from the ski-ramps on the RN's carriers. In early 1997, No. 1 Squadron deployed four GR7s aboard HMS *Illustrious* as part of Operation Ocean Wave '97. This marked the first time that FA2 Sea Harriers and GR7 Harrier IIs operated as a combined force, where the Sea Harriers mainly provided combat air patrol and the GR7s formed the strike force. GR7s receive limited modifications prior to boarding carriers, including the addition of a new I-band IFF transponder.

In 1989 BAe proposed a Harrier GR9 that used the Blue Vixen radar and was equipped with AIM-120 missiles, along with the uprated Pegasus 11-61 engine. This would give capabilities similar to the American AV-8B+ variant. Funding constraints make it very unlikely this variant will ever be approved.

The GR7 is the ultimate, and probably last, RAF version of the Harrier. Noteworthy is the curved leading edge on the LERX—later USMC Harriers have a straight leading edge, and it is better faired into the upper surface of the wing. (Peter Steehouwer)

Although unguided rockets have largely lost their appeal in the era of precision-guided munitions, the Harrier is still capable of carrying them, and routinely appear at airshows with various rocket pods slung under the wings. Like their American counterparts, the GR7s can carry a retractable in-flight refueling probe on the left air intake. (Nick Challoner)

GR7 with a typical weapons load – BL755 cluster bombs and AIM-9 Sidewinders. (British Aerospace)

Two-Seat Harrier IIs

The RAF continued to use the existing T4A two-seaters to train pilots for the GR5. The advent of the GR7 forced the RAF to consider modifying a small number of the T4s with the night attack avionics fitted to the GR7, to become T6s. On 28 February 1990 the RAF ordered 14 two-seat versions of the Harrier II designated T10, at a total cost of £200 million. This total was later reduced by one aircraft in order to compensate for a change in the currency exchange rate. At the same time the plan to convert the T4s to the T6 configuration was abandoned.

Unlike the TAV-8B, the T10 two-seater is fully mission-capable, and includes a complete set of weapons stations under the wing, as well as a FLIR in the nose. (Nick Challoner)

The T10 uses the same airframe as the TAV-8B with the addition of most GR7 systems. The first T10 was assembled at Dunsfold as all other RAF Harriers had been, but subsequent T10s were assembled at Warton since the Dunsfold facility was busy modifying Sea Harries to the FA2 configuration. The first T10 (ZH653) made its maiden flight on 7 April 1994, and the first operational aircraft (ZH657) was delivered to the RAF on 30 January 1995. Unlike the TAV-8B, the T10 is fully capable of operational missions, and carries the full complement of eight underwing weapons stations. No. 20 (Reserve) Squadron received ten T10s, and each operational squadron received a single example.

The T10s are, by design, somewhat heavier than the GR7, and suffer from marginal hover performance, especially in warmer climates. There have been discussion about fitting the aircraft with the Pegasus 11-61 engine used on the Italian two-seaters, but funding limitations have kept this from occurring.

The UK's 1998 strategic defense review (SDR) contained a few surprises. Amongst these was the call to build two new 40,000-ton aircraft carriers to replace the three "command cruisers" currently in service. Each of the new ships will carry a mix of 50 fixed-wing aircraft and helicopters. Initially the fixed wing aircraft will be the existing Harriers, although they are expected to be replaced by new aircraft around 2010.

The second recommendation that directly affected the Harrier force was a call to combine the Harrier assets of the RAF and RN into a joint RAF/RN fixed-wing STOL force. There are no indications that this will drive more commonality between the two current types of Harriers—GR7 and FA2—since funding will continue to be limited. Plans are to replace all the Harriers with a common aircraft type in the future. This is currently envisioned as the JSF, but could potentially include derivatives of the Rafale or Eurofighter.

A good view at the eight pylon wing used on the GR5 and GR7. American AV-8Bs do not have the pylon immediately ahead of the outrigger landing gear. Notice the tight clearance between the centerline store and the cannon pods. (British Aerospace)

Spanish and Italian Harriers

Spain (*Arma Aérea de la Armada*) had purchased AV-8As through the United States during a time when their relations with Britain were seriously strained. Since the aircraft were operated from aircraft carriers, it was widely assumed that any follow-on purchase would consist of Sea Harriers. Nevertheless, Spain placed an order for 12 EAV-8Bs in March 1983, designating the new aircraft VA.2 Matador. The first 3 aircraft (BuNo 163010/012, VA.2-1/-3) were ferried to Rota on 6 October 1987, with the remaining aircraft delivered by the end of 1989.

Showing that most of the previous hard feelings with the British had been dismissed, the Spanish asked BAe to assist in qualifying the new aircraft for operation from the 12° ski-ramp on the *Principe de Asturias*. The Spanish Navy also rented time in the RAF GR3 simulator instead of sending pilots all the way to Cherry Point. After Spain received their own AV-8B simulator, the RAF rented time in it to train pilots pending delivery of their own GR5 simulator.

Like the Marine Corps, Spain opted to use the six hard-point wing instead of the RAF's eight. This means that the outboard weapons station is almost always dedicated to carrying an AIM-9 Sidewinder. Like the Americans and Italians, the Spanish opted to use the new GAU-12 25-mm cannon instead of the RAF's Adens. (Boeing)

Nine new-build EAV-8B+ aircraft (BuNo 165028/035, VA.3-2/10) were procured by the Spanish Navy in 1990, and the remaining EAV-8Bs will be brought up to the same standard as funding permits. A single two-seater TAV-8B+ (BuNo 165027, VA.3-1) was also ordered.

The Italian Navy *(Marina Militare Italia)* was prohibited by a 1937 law from operating fixed-wing aircraft. The only reason they could operate

Spain was the first county other than the US and UK to order the Harrier II. The first VA.2 Matadors were essentially identical to the early AV-8B configuration, although they are being brought up to AV-8B+ standards. Spain also ordered a small number of new-build AV-8B+ aircraft. The forward LID under the fuselage between the strakes shows up well here. These three LIDs (the forward dam, plus the two strakes, or gun pods) help to form a box that captures air during hover, creating additional lift. (Nick Challoner)

The Italian Harrier II+ is identical to the USMC version, including the ALE-39 countermeasure dispensers on the upper rear fuselage. Italy purchased the aircraft with the understanding that they were AIM-120 AMRAAM capable, but qualification of this weapon has been slow due to budget issues in the US. (via Jan Michels)

The second Italian two-seater. (Andy Moulton via Kev Darling/BBA)

The first Italian two-seater. Italy was the first country to order the trainer version ahead of single-seater. In this case, it worked well since Italy is skipping the early versions of the AV-8B and going directly to the radar-equipped AV-8B+. Although there appears to be a radome on the trainer, the aircraft does not carry a radar. (Nick Challoner)

helicopters was because the technology was unknown when the law was drafted. All aircraft were to be operated by the Air Force, although the Navy was in charge of tasking antisubmarine and antiship operations. Nevertheless, the MMI began looking at the Harrier in October 1967 after Hawker pilot Hugh Merewethere landed a Harrier on the anti-submarine carrier *Andrea Doria*. A proposal was made for an offset deal involving six British-built Mk 50s along with kits for 44 aircraft to be assembled in Italy. In exchange the UK would purchase Italian-built Chinook helicopters. Controversy within the Italian military eventually killed the idea.

However, when the new helicopter carrier *Geuseppe* was launched in 1983, it was equipped with a 6.5° ski-ramp, clearly an indication the Italian Navy wanted to operate Harrier-type aircraft. The ramp was explained as a device to keep excessive sea spray from swamping the flight deck – the ruse fooled nobody. After a prolonged evaluation of the Sea Harrier and AV-8B, the MMI obtained government approval to operate fixed-wing aircraft in 1989. In May 1989 two TAV-8B two-seaters (BuNo 164136/137, MM55032/33) were ordered from McDonnell Douglas, and a small cadre of pilots was dispatched to the United States for training.

Sixteen AV-8B+ single-seaters were ordered on 24 April 1994. The first three (BuNo 164563/565, MM7199/7201) were diverted from a Marine Corps order and delivered to MCAS Cherry Point for training. In 1994 the 2 two-seat aircraft were reengined with F402-RR-408s to improve hot weather performance. The remaining 13 single-seaters (BuNo 156007/019, MM7212/7224) were delivered in kit form to Alenia, while some offset work was given to Aerea (pylons), Breda (cannon), Fiat Avia (engine components), and Magnaghi (hydraulics).

The Italian Harriers have already seen combat, providing fire support during the evacuation of UN troops from Somalia in January 1995. Over 100 combat missions were flown using mainly rocket pods and 25-mm cannon.

The Italian aircraft are essentially identical to the equivalent USMC aircraft, although the addition of the -408 engine to the two-seaters makes them arguably the best of the type.

A nice illustration of the differences between the Marine AV-8B+ (top) and the RAF GR7 (bottom). Noteworthy is the difference in size and shape of the LERX, with the British aircraft using the so-called 75% design, while the American is using the later 100% unit. The extra weapons station on the GR7 is carrying a Sidewinder. The RAF has not adopted the American ALE-39 chaff/flare dispenser on the rear fuselage, which has resulted in the ram air intake at the base of the vertical stabilizer retaining its original shape instead of being extended forward. The new nose profile on the AV-8B+ to accommodate the radar unit is also obvious. For two aircraft built on essentially the same assembly line, they differ considerably in detail. (British Aerospace via Jan Michels)

SIGNIFICANT DATES

KEY DATES IN THE HISTORY OF THE HARRIER

1483
Leonardo da Vinci makes the first sketch of a helicopter

1935
First successful coaxial-rotor helicopter flies in France

1939
First successful single-rotor helicopter flies in the United States

April 1947
The US Navy contracts with Ryan Aeronautical to investigate jet-powered VTOL flight

Early 1950s
French engineer Michel Wibault develops a vectored-thrust engine concept

31 May 1951
Ryan conducts the first free-flight hover by a jet-powered vehicle

28 May 1956
Ryan X-13 completes first vertical test flight

January 1955
Bell Aircraft submits four reports urging the development of a VTOL fighter

July 1955
Bell receives contract to develop X-14 test aircraft

Early 1956
Wibault presents his engine concept to the French and American defense departments. Neither is interested

Later in 1956
Ryan conducts the first free-flight hover by a jet-powered vehicle

17 February 1957
Wibault presents to the NATO MWDA, and finds minor interest

Early 1957
Bristol completes preliminary improvements to Wibault's design

1957
Bristol and Wibault patent what would become the Pegasus engine

Late 1959
Hawker receives a contract for two P.1127 test aircraft

December 1959
NASA begins wind tunnel tests on the P.1127

21 October 1960
First P.1127 hovers

13 March 1961
First P.1127 makes first conventional flight

14 December 1961
Bill Bedford ejects from P.1127 during first accident

Early 1962
British, American, and German governments purchase nine Kestrels

8 February 1963
A P.1127 lands aboard HMS *Ark Royal*

7 March 1964
First Kestrel flies

15 October 1964
Tripartite Squadron officially formed

January 1965
Hawker proposes two-seat P.1127

31 August 1966
First preproduction Harrier makes its maiden flight

22 April 1969
T2 two-seater makes its first flight

5 May 1969
Harrier from No. 1 Squadron participates in the Daily Mail Transatlantic Air Race, and wins one leg

1968
US Marine Corps orders 12 Harriers

1972
Spain orders Harriers – via the US Marine Corps

13 December 1973
26-volume report issued on advanced Harrier derivatives for US Marine Corps and Royal Air Force

Early 1975
UK withdraws from big-wing Harrier development project

27 July 1976
US approves development of the big-wing AV-8B

20 August 1978
First Sea Harrier flies

9 November 1979
YAV-8B prototype flies

December 1979
India orders Sea Harriers

11 March 1982
A Pegasus engine equipped with digital engine controls flies for the first time in a modified GR3

April 1982
Harriers and Sea Harriers see combat in Falkland Islands

29 August 1983
First pilot-production AV-8B flies

30 April 1985
First DB GR5 flies

Late 1985
AV-8B(NA) development is authorized

21 October 1986
First TAV-8B two-seater flies

1988
BAe and McDonnell Douglas announce development of a radar-equipped version of the Harrier II

May 1989
Italy orders 2 two-seat Harrier IIs

1991
USMC Harriers fly 3,380 missions during Operation Desert Storm, resulting in the loss of 5 aircraft

January 1995
Italian Harriers enter combat over Somalia in support of UN forces

30 August 1995
RAF GR7s enter combat over Bosnia in support of UN forces

Acronyms

Deciphering the Alphabet Soup

ALARM
Air-Launched Anti-Radar Missile (UK)

AMRAAM
Advanced Medium Range Air-to-Air Missile (AIM-120)

ATV
Air Test Vehicle (Ryan)

AWACS
Airborne Warning and Control System (E-3 Sentry)

BAe
British Aerospace

DB
Development Batch (equal to US 'Y' aircraft)

DoD
Department of Defense (US)

ECM
Electronic Countermeasures

FLIR
Forward-Looking Infrared

FM
Frequency Modulation (radio)

FSD
Full-Scale Development

HUD
Heads-Up Display

I/WAC
Interface/Weapon Aiming Computer

IFF
Identification, Friend or Foe

JSF
Joint Strike Fighter

lbf
pounds force (Thrust)

LERX
Leading Edge Root Extension

LID
Lift Improvement Devices

LRMTS
Laser Ranging and Marked-Target Seeking (System)

M.o.S.
Ministry of Supply (UK)

McAir
McDonnell Douglas Corporation

MCAS
Marine Corps Air Station

MFD
Multifunction Display

MMI
Marina Militare Italia (Italian Navy)

MWDA
Mutual Weapons Development Agency

NA
Night Attack

NASA
National Aeronautics and Space Administration

NATO
North Atlantic Treaty Organization

NVG
Night Vision Goggles

OBOGS
On-Board Oxygen Generation System

OCU
Operational Conversion Unit

OSCAR
Open Systems Core Avionics Requirements

PCB
Plenum Chamber Burning

RAF
Royal Air Force (UK)

RN
Royal Navy (UK)

RNAS
Royal Navy Air Station

RWR
Radar Warning Receiver

SAS
Stability Augmentation System

STO
Short Takeoff

STOL
Short Takeoff and Landing

UHF
Ultra High Frequency (radio)

UK
United Kingdom

US
United States

USMC
United States Marine Corps

USS
United States Ship

V/STOL
Vertical / Short Takeoff and Landing

VATOL
Vertical Attitude Takeoff and Landing

VHF
Very High Frequency (radio)

VTO
Vertical Takeoff